Introduction to Construction Project Engineering

WITHDRAWN

This new textbook fills an important gap in the existing literature, in that it prepares construction engineering and built environment students for their first experience of the jobsite. This innovative book integrates conceptual and hands-on knowledge of project engineering to introduce students to the construction process and familiarize them with the procedures and activities they need to operate as project engineers during their summer internships and immediately after graduation.

The textbook is structured into four sections:

- Section A: Introductory Concepts
- Section B: Field Engineering
- Section C: Office Engineering
- Section D: Advanced Project Engineering

The emphasis on field tasks and case studies, questions, and exercises taken from across civil works and commercial building sectors makes this the ideal textbook for introductory to intermediate courses in Construction Engineering, Construction Engineering Technology, Civil and Architectural Engineering, and Construction Management degree programs.

Giovanni C. Migliaccio is an Associate Professor in Construction Management at the University of Washington (UW), USA. He is also the Associate Director for the UW Center for Education and Research in Construction, holds a P.D. Koon Endowed Professorship and an affiliate fellowship with the UW Runstad Center for Real Estate Studies.

Len Holm is a Senior Lecturer in Construction Management at the University of Washington, USA, and a Construction Management Professional.

D0144367

Introduction to Construction Project Engineering

Giovanni C. Migliaccio and Len Holm

Routledge
Taylor & Francis Group

LONDON AND NEW YORK

First published 2018
by Routledge
2 Park Square, Milton Park, Abingdon, Oxon OX14 4RN

and by Routledge
711 Third Avenue, New York, NY 10017

Routledge is an imprint of the Taylor & Francis Group, an informa business

British Library Cataloguing-in-Publication Data
A catalogue record for this book is available from the British Library

Library of Congress Cataloging-in-Publication Data
Names: Migliaccio, Giovanni C. (Giovanni Ciro), 1968– author. | Holm, Len, author.
Title: Introduction to construction project engineering / Giovanni C. Migliaccio and Len Holm.
Description: Abingdon, Oxon : Routledge, 2018. | Includes bibliographical references and index.
Identifiers: LCCN 2017047488 | ISBN 9781138736559 (hardback : alk. paper) |
ISBN 9781138736580 (pbk. : alk. paper) | ISBN 9781315185811 (ebook)
Subjects: LCSH: Building–Superintendence. | Construction projects.
Classification: LCC TH438 .M497 2018 | DDC 690.068–dc23
LC record available at https://lccn.loc.gov/2017047488

ISBN: 978-1-138-73655-9 (hbk)
ISBN: 978-1-138-73658-0 (pbk)
ISBN: 978-1-315-18581-1 (ebk)

Typeset in Bembo
by Out of House Publishing

To my wife, Tatiana Paola, for her love and support throughout the writing of this book. To my parents, Giuseppe and Adriana for instilling in me a drive to learn. To my co-author Len, for motivating me in pursuing this writing endeavor since its conception.

G.C.M.

To my wife, Jane, for being my partner in writing, work, life, and love. To my father who taught me at the age of 10 how to swing a hammer and gave me the opportunity to do something in construction that he had never had the chance to do. And to our children who are already carrying on their grandfather's legacy.

A.L.H.

Contents

Preface

This book could actually have been titled "Bridging the Gap" or, as we refined our proposal further, "Bridging Several Gaps." Throughout our academic careers, we have taught numerous construction management courses at the undergraduate and graduate levels, including *Introduction to Construction Management*, and we both have construction industry experience. One of us has over 40 years of experience as a construction practitioner and has often directly supervised recent graduates from college programs in construction. We felt there was a gap in the available textbook coverage, in that most books geared for an introduction to a construction management course are likely titled *Introduction to Project Management*. Students just embarking on their construction management or construction engineering education need to be provided with an introduction to the construction industry, particularly as it relates to early internships and job opportunities after graduation, for construction field engineers or project engineers. Most of these students will not be achieving the position of project manager for five to seven years, and some of them never will have that title. We therefore structured this book as an *introduction to the construction industry*, specifically targeted *for the project engineer*.

Another gap we feel exists in university construction education is for architectural and civil engineers. Most existing introductory books are written from the client or architect's perspective. Those which feature construction applications only utilize commercial construction case studies. These books are difficult to adapt to engineering programs, most of which have an *Introductory Construction* course. Therefore, we have also bridged this gap by developing a heavy-civil case study project and weaving examples from that project, along with a mixed-use case study project, throughout the book. Descriptions of both of our cases are included in Appendix A, and example documents are included in many of the chapters. Review questions and advanced application exercises, including math problems, are included in each chapter and connect to these cases as well. Additional case study backup, along with answers to the review questions, is included on a companion website (www.routledge.com/cw/migliaccio). The projects, companies, and participants described in our case studies are fictitious, but the detailed estimates, schedules, and example documents are based upon similar actual projects that the authors participated in professionally. Additional appendices include the Abbreviations, Glossary, and Index.

Each of our chapters features specific applications of construction management principles and tools for the project engineer. Early chapters begin with a basic foundation in construction management and entry-level field and project engineering responsibilities. Each chapter builds upon its predecessors, similarly to a contractor building upon a structure's foundations and superstructure. Responsibilities increase for the project engineer, eventually culminating with advanced project engineering applications, including risk management and our final *Introduction to construction project management* chapter.

We would like to thank estimating instructor and industry professional Larry Bjork, Associate Professors Ken Yu Lin and Abdel Aziz, and civil engineering Associate Professor Steve Muench for their advice and review of chapter drafts. Dean and Professor John Schaufelberger provided us with early input and mentoring, and his words of wisdom are always welcome. Jane Holm contributed an early draft of Chapter 13, and we appreciate her passion for sustainability. We want to also thank Kel Mejlaender for helping us with proofreading, editing, and streamlining all the chapters in this book. Dr. Migliaccio had the opportunity to work for PCL Construction on three different projects during the writing of this book, and his exposure to applied jobsite project engineering applications was very beneficial. Matt Glassman at PCL Construction provided several figures for this book. Lastly, we would like to thank the University of Washington construction management students, who were able to use drafts of this material in the classroom and provide us with real-time feedback.

We hope you enjoy the material and what we feel is a unique approach to an *Introduction to Construction Project Engineering*. Please feel free to contact us with suggestions for future editions—did we bridge the gap?

Giovanni Migliaccio (gianciro@uw.edu) and
Len Holm (holmcon@aol.com)

Section A

Introductory concepts

1 Introduction

The built environment

Several definitions of "Built Environment" (BE) exist. Some definitions can be quite broad to include essentially anything humankind creates that is not living, such as automobiles, planes, buildings, infrastructures, trains, and even shoes and clothing. Throughout this book, a common definition of the built environment will be used to include facilities and physical infrastructures that add or change functions in the underlying natural, economic, and social environments. Therefore, the focus is limited to structures we design and build. This includes buildings for people to live and work in, roads and bridges for cars to drive on, airports and airplane hangars for planes to operate from, railroad tracks, tunnels, and trestles for trains to use, and monuments for people to admire. Thus, shoes and clothing are not considered part of the built environment, but the factory they were produced in is part of the built environment. Paintings and sculptures are not part of the built environment, but the museums they are hosted in and the public spaces they are placed in are part of the built environment.

Built environment industries

Developing the built environment—adding or changing functions in the underlying natural, economic, and social environments—means first *envisioning* facilities and physical infrastructures. These facilities and infrastructures have residential, commercial, governmental, industrial, recreational, transportation, and utility uses. Once envisioned, the next step is *building* them. These tasks are multi-faceted and highly specialized, so several types of businesses exist that support their clients to help develop the built environment. In categorizing these businesses by industry, this book refers to the North American Industry Classification System (NAICS), which "is the standard used by federal statistical agencies in classifying business establishments for the purpose of collecting, analyzing, and publishing statistical data related to the U.S. business economy" (NAICS 2017). An essential BE industry includes companies within the *Architectural, Engineering, and Related Services* NAICS category, which provide planning, design, and other consulting services. It is these services that institutions, clients, and contractors require to envision built environment undertakings. The title of this book includes the word "construction," because companies within the *Construction* category are those physically translating design into facilities and infrastructures. Together, companies within the *Architectural, Engineering, and Related Services* and *Construction* NAICS categories directly contribute to the economic output of a nation, while indirectly supporting other industries. Given their built environment-centered focus, these companies are sometimes referred to as a single industry: the architecture–engineering–construction industry (AEC). The following sections provide a current overview of these industries through critical statistics from the United States (U.S.) Bureau of Labor Statistics (BLS) and the U.S. Bureau of Economic Analysis (BEA).

Architectural, engineering, and related services (AE)

AE businesses provide a large number of services that are needed to develop the built environment, including architectural, landscape architectural, engineering, drafting, surveying and mapping, and testing services. According to the BEA data, the gross output of the AE industry peaked in 2008 at $285 billion. Though the AE industry was hit by the Great Recession, its economic performance has slowly recovered since 2011. In 2015, its gross output was equal to $280 billion, nearing its 2008 peak.

Construction

Construction businesses provide many services that are needed to develop the built environment, including the construction of buildings and infrastructures. Construction services may contribute to the development of new facilities/infrastructures, as well as additions, alterations, repairs, or maintenance to existing facilities/infrastructures. NAICS subdivides construction businesses into three sectors: (a) Construction of Buildings, (b) Heavy and Civil Engineering Construction, and (c) Specialty Trade Contractors.

According to the BEA data, the dollar value of the construction industry reached a peak in 2006 when it added $698 billion to the U.S. economy, which was equal to 5.0% of the gross domestic product (GDP) of the United States, with a gross output of $1,345 billion. While the industry was hit hard by the Great Recession between 2007 and 2010, its economic performance has improved again since 2011. In 2016, its dollar value added $784 billion to the U.S. economy, which was equal to 4.2% of the GDP, with a gross output of $1,433 billion. Whereas these economic statistics are massive in absolute numbers, the contribution of the construction industry to the GDP has been fluctuating from 3.5% to 5% since 1947.

The built environment by functional role and occupation

There are many different agencies, companies, and people involved in the built environment. The project owner, also simply referred to as the owner or the client, is the company or individual at the top of the organization chart. The project owner will have in-house representatives who contractors and designers communicate with and report to. Some larger owner companies may also have in-house design and construction capabilities. The owner's facility manager and maintenance personnel take over operations of the building once the construction is complete.

The design team is headed by an architectural firm for commercial projects and by a civil engineering firm for heavy-civil projects. These firms will have a contact person, who may carry the title of project manager (PM), similar to that of a contractor. The lead design firm also employs a variety of other design disciplines, such as landscape architecture, structural engineering, mechanical engineering, and electrical engineering. There are a variety of other firms involved in the development of built environment projects that may contract directly with the owner as consultants or serve the lead design firm as sub-consultants in various specialties, including estimating, scheduling, lighting, and waterproofing. Each of these consulting firms will have a variety of individuals working for them in addition to their PM, including the architect, lead designer, engineers, draftsmen, computer-aided design (CAD) or building information modeling (BIM) technicians, specification writers, and shop drawing checkers.

Employees in the architecture and engineering disciplines are a significant portion of the AEC industry, as shown by its employment data. Table 1.1 shows the most recent employment and wage estimates for some of the architecture and engineering occupations that support the AEC industry.

Table 1.1 2016 Employment and Wage Estimates for Selected Architecture and Engineering Occupations

Occupation Title	Employment	Mean Hourly Wage	Annual Mean Wage
All Occupations	**140,400,040**	**$23.86**	**$49,630**
Architecture and Engineering Occupations (All)	*2,499,050*	*$40.53*	*$84,300*
Architects, Surveyors, and Cartographers	*174,720*	*$36.66*	*$76,260*
Licensed Architects (★)	99,860	$40.61	$84,470
Landscape Architects	19,420	$33.08	$68,820
Surveyors	43,340	$30.52	$63,480
Engineers (All)	*1,635,420*	*$46.37*	*$96,440*
Civil Engineers	287,800	$43.14	$89,730
Electrical Engineers	183,770	$47.41	$98,620
Environmental Engineers	52,280	$42.56	$88,530
Health and Safety Engineers (★★)	25,410	$43.36	$90,190
Mechanical Engineers	285,790	$43.17	$89,800
Drafters and Technicians (★★★)	*688,900*	*$27.66*	*$57,530*
Architectural and Civil Drafters	96,810	$26.10	$54,290
Civil Engineering Technicians	72,150	$25.06	$52,120
Environmental Engineering Technicians	16,550	$25.24	$52,500
Surveying and Mapping Technicians	53,920	$21.87	$45,490

(★) Except Landscape and Naval Architects
(★★) Except Mining Safety Engineers and Inspectors
(★★★) Drafters, Engineering Technicians, and Mapping Technicians

Adapted from U.S. Bureau of Labor Statistics, May 2016 National Occupational Employment and Wage Estimates, United States (BLS 2017)

Based on these estimates, architecture and engineering occupations employ nearly 2.5 million individuals, equal to 2% of the domestic workforce. However, a large number of these individuals work for owner organizations or other entities that support the development of the built environment, such as governmental regulatory agencies.

There are a variety of firms that fit under the title of contractors. The prime contractor is typically the firm that contracts with the owner and, in turn, employs subcontractors and suppliers. A construction manager (CM) or a general contractor (GC) usually serves as the prime contractor on a project. Specialty contractors are firms that focus on one type of work, such as electrical, whereas the GC is responsible for all of the work. Specialty contractors usually, but not necessarily, participate in a project as subcontractors. Most of these firms will be introduced in Chapter 3. Subcontractors and suppliers are also the focus of Chapter 19.

Every general or specialty contractor customarily employs PMs, superintendents, project engineers, foremen, and craftsmen. Construction management firms usually do not employ foremen or craftsmen, but they will have PMs, superintendents, and project engineers. Every large construction firm will have hundreds, if not thousands, of employees who fit these and other titles. However, a smaller construction firm is often led by its proprietor, who needs to multi-task. One of the authors' fathers was a carpenter and a self-employed master-builder GC, who wore many hats at the same time, including those of accountant and estimator.

Employees of construction firms are the largest portion of the AEC industry, as shown by its employment data. Table 1.2 shows the most recent employment and wage estimates for some of the construction occupations that support the BE industry. Based on these estimates, construction occupations employ nearly 5 million individuals, equal to 4% of the domestic workforce. Figures 1.1 and 1.2 show employment and wage estimates by construction trade.

Table 1.2 2016 Employment and Wage Estimates for Selected Construction Occupations

Occupation Title	Employment	Mean Hourly Wage	Annual Mean Wage
All Occupations	**140,400,040**	**$23.86**	**$49,630**
Construction Managers (★)(★★)	249,650	$47.84	$99,510
First-Line Supervisors of Construction Trades and Extraction Workers (★★★)	538,220	$32.71	$68,040
Construction Trades Workers (★★)	4,216,890	$22.88	$47,580
Helpers, Construction Trades	228,590	$14.86	$30,900
Other Construction and Related Workers	403,940	$22.92	$47,670

(★) Including all individuals who "Plan, direct, or coordinate, usually through subordinate supervisory personnel, activities concerned with the construction and maintenance of structures, facilities, and systems." This is a broad description that could include project managers, project superintendents, and project engineers.

(★★) Excluding first-line supervisors.

(★★★) Including all individuals who "Directly supervise and coordinate activities of construction or extraction workers." This statistic is cumulative and the data do not allow the segregation of supervisors of construction trades from supervisors of extraction workers.

Adapted from U.S. Bureau of Labor Statistics, May 2016 National Occupational Employment and Wage Estimates, United States (BLS 2017)

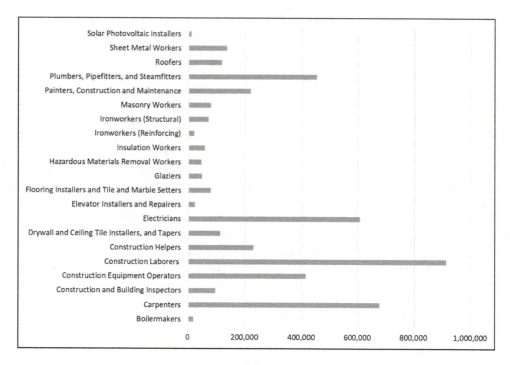

Figure 1.1 Construction Employment by Trade (May 2016)
Adapted from U.S. Bureau of Labor Statistics, May 2016 National Occupational Employment and Wage Estimates, United States (BLS 2017)

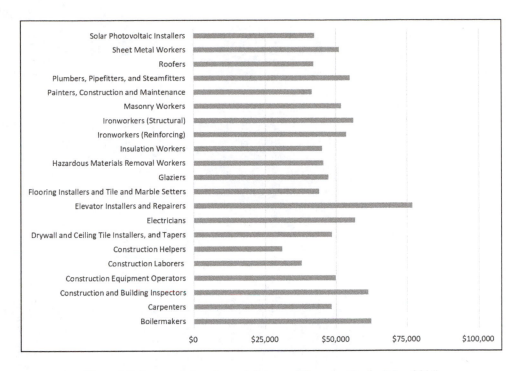

Figure 1.2 Construction Annual Average Wages by Trade (May 2016)
Adapted from U.S. Bureau of Labor Statistics, May 2016 National Occupational Employment and
Wage Estimates, United States (BLS 2017)

There are many different specialized trades or crafts involved in construction, from plumbers to ironworkers to electricians, among others. Some of these trades are included in the categories listed in Table 1.2. They are all experts in their focus area and it is a gross misrepresentation to call them all "laborers" or simply "workers." Instead, their mastery in a construction trade should be recognized by correctly identifying and distinguishing one trade from another. Table 1.3 lists and describes major construction occupations by trade and provides employment and wage estimates as of May 2016.

The built environment by project type

There are many differences between the AEC industry and other industries, such as automotive manufacturing. Foremost, AEC products are built projects. Each project is unique and is built on a separate site at a different time, under different weather conditions, with different design parameters and different design and construction teams, especially at the craftsman level. Cars are instead made in a factory's assembly line. The exact same car is mass-produced, over and over again.

The built environment can be subdivided or classified in a variety of fashions. One of the most common classifications of the built environment is by use:

- *Residential building projects* include single-family homes, apartments, condominiums, and retirement facilities. These projects usually consist of low- or mid-rise buildings that rely on

Table 1.3 Major Construction Occupations by Trade

Occupation	Job Summary	Employees	2016 Mean Annual Pay
Boilermakers	Boilermakers assemble, install, and repair boilers, closed vats, and other large vessels or containers that hold liquids and gases.	16,600	$62,200
Carpenters	Carpenters construct and repair building frameworks and structures—such as stairways, doorframes, partitions, rafters, and bridge supports—made from wood and other materials. They also may install kitchen cabinets, siding, and drywall.	676,980	$48,340
Construction and Building Inspectors	Construction and building inspectors ensure that construction meets local and national building codes and ordinances, zoning regulations, and contract specifications.	94,960	$61,250
Construction Equipment Operators	Construction equipment operators drive, maneuver, or control the heavy machinery used to construct roads, bridges, buildings, and other structures.	412,190	$49,810
Construction Laborers and Helpers	Construction laborers and helpers perform many tasks that require physical labor on construction sites.	Laborers: 912,100 Helpers: 228,590	$37,890 $30,900
Drywall and Ceiling Tile Installers, and Tapers	Drywall and ceiling tile installers hang wallboard and install ceiling tile inside buildings. Tapers prepare the wallboard for painting, using tape and other materials. Many workers both install and tape wallboard.	111,650	$48,460
Electricians	Electricians install, maintain, and repair electrical power, communications, lighting, and control systems in homes, businesses, and factories.	607,120	$56,650
Elevator Installers and Repairers	Elevator installers and repairers install, fix, and maintain elevators, escalators, moving walkways, and other lifts.	22,240	$76,860
Flooring Installers and Tile and Marble Setters	Flooring installers and tile and marble setters lay and finish carpet, wood, vinyl, and tile.	77,410	$43,950
Glaziers	Glaziers install glass in windows, skylights, and other fixtures in storefronts and buildings.	47,140	$47,260
Hazardous Materials Removal Workers	Hazardous materials (hazmat) removal workers identify and dispose of asbestos, lead, radioactive waste, and other hazardous materials. They also neutralize and clean up materials that are flammable, corrosive, or toxic.	44,280	$45,500

Occupation	Description	Employment	Median Salary
Insulation Workers	Insulation workers install and replace the materials used to insulate buildings to help control and maintain the temperatures in buildings.	56,770	$45,070
Ironworkers	Ironworkers install structural and reinforcing iron and steel to form and support buildings, bridges, and roads. Reinforcing: Structural:	20,020 69,440	$53,600 $56,040
Masonry Workers	Masonry workers, also known as *masons*, use bricks, concrete blocks, concrete, and natural and manmade stones to build walls, walkways, fences, and other masonry structures.	77,560	$51,770
Painters, Construction and Maintenance	Painters apply paint, stain, and coatings to walls and ceilings, buildings, bridges, and other structures.	217,280	$41,510
Plumbers, Pipefitters, and Steamfitters	Plumbers, pipefitters, and steamfitters install and repair pipes that carry liquids or gases to, from, and within businesses, homes, and factories.	451,500	$54,870
Roofers	Roofers replace, repair, and install the roofs of buildings using a variety of materials, including shingles, bitumen, and metal.	116,410	$42,080
Sheet Metal Workers	Sheet metal workers fabricate or install products that are made from thin metal sheets, such as ducts used in heating and air conditioning systems.	134,450	$51,080
Solar Photovoltaic Installers	Solar photovoltaic (PV) installers, often called *PV installers*, assemble, install, or maintain solar panel systems on roofs or other structures.	8,870	$42,500

Adapted from U.S. Bureau of Labor Statistics, Construction and Extraction Occupations, United States (BLS 2015) mainly light-frame construction. *High-rise condo* and *apartment buildings* have a primarily residential focus. However, these projects are often considered commercial due to their reliance on construction techniques and materials similar to those of commercial and government projects.

- *Commercial and governmental building projects* include building projects for uses other than residential or heavy industrial. These projects may consist of low-, mid-, or high-rise buildings that rely on a more specialized spectrum of construction techniques and materials than light-frame construction. Examples include offices, fire stations, libraries, retail stores, and schools. Given the breadth, this sector is often subdivided into subsets. For instance, *hospitality projects* are a subset of commercial projects in terms of construction techniques and materials but with a narrow focus on hospitality, such as movie theaters, restaurants, hotels (sometimes developed with construction techniques and materials similar to those of residential projects), and golf courses (sometimes developed with construction techniques and materials similar to those of heavy-civil projects). Similarly, *healthcare projects* are another commercial subset with a stronger focus on building technologies and systems. *Mixed-use projects* include crossover building projects that include several different uses under the same roof. An extreme example would be a downtown high-rise building hosting below-grade parking and a bowling alley, a retail space at street level, a restaurant on the second floor, a boutique hotel on the next five floors, condominiums on the ten floors above, and a swimming pool on the roof. A simpler example of a mixed-use project is the subject of a case study we will use throughout this textbook and briefly described in Appendix A.
- *Heavy-civil or infrastructure projects* include roads, bridges, tunnels, dams, and the distribution lines, related buildings, and structures for utilities (i.e., water, sewerage, petroleum, gas, power, and communication). One of our case studies for this book is a heavy-civil highway overpass project, and it is briefly described in Appendix A.
- *Heavy-industrial projects* include, but are not limited to, power plants, chemical plants, and oil refineries. Since these projects include facilities that are process-oriented, their design and construction is highly dependent upon the specific process to be carried out in the facility.

Book overview

This book is organized into four major sections, A–D. The content and learning objectives of each section are described below.

Section A – Introductory concepts

In addition to this introductory Chapter 1, Section A features five additional chapters, which provide the reader with a basic understanding of construction management topics, which the rest of the book relies upon. Chapter 2 expands upon the types of built environment projects introduced above. The major phases that each project experiences from planning, design, and pre-construction, through construction, close-out, commissioning, and operations are discussed. Additionally, the impacts of different design decisions on projects are evaluated against each of these phases.

As discussed above, there are many different participants in construction, including project owners, designers, and contractors. All of these project participants are elaborated on in Chapter 3, along with different corporate structures. Chapter 4 introduces the reader to project delivery, beginning with the traditional design–bid–build (DBB) method and the more recent construction manager/general contractor (CM/GC), design–bid (DB), and integrated project delivery (IPD) methods. A comparison of public versus private clients and their preferred delivery processes is also explored. Procurement selection methods and types of contracts are the focus of Chapter 5.

Chapter 6 is the introduction to the rest of the book. A brief introduction to estimating, scheduling, and project controls provides the foundation for several subsequent more detailed chapters on each of these topics. The role of the project engineer (PE) is discussed in each of these chapters and throughout the book. Each section builds upon the others and adds more advanced and technical tasks and responsibilities for the PE. Each chapter also narrates the latest technological tools available to a PE in the context of the chapter topic.

Section B – Field engineering

We have introduced the terms "field," "office," and "advanced project" engineering, along with the customary "project" engineering, in this book. Many companies use the terms interchangeably. We are not making any hardline distinctions here; rather, we are using the terms loosely for areas the project engineer may work on or types of activities he or she may be tasked with. We will generically use the term "project" engineering for all positions and responsibilities throughout the book. In Section B, we focus on the entry-level project engineer who works in the field or jobsite trailer and may assist the project manager, project superintendent, or senior project engineer with their duties and responsibilities to achieve project success, in terms of cost, time, quality, and safety. All of these project controls are important, but a construction project that meets its schedule and cost goals and is of acceptable quality is not truly a successful project if someone is seriously injured. Safety control is the focus of Chapter 7 and discusses a variety of safety tools and documentations involving workers' compensation insurance, drug and alcohol testing, hazardous material planning, and heavy-civil equipment management.

Production control, or cost and schedule control of the field, is primarily the responsibility of the superintendent. The field or project engineer will support the superintendent with cost coding, development and maintenance of work packages, and schedule monitoring. Chapter 8 discusses these topics and some of the PE's production control responsibilities. Chapter 9 revolves around a series of activities that we label as "active" quality control (QC) and monitoring; these are project-specific efforts and not generic or boilerplate. The PE assists the superintendent with active QC activities, such as subcontractor coordination, third-party testing and inspection, and QC documentation.

Section C – Office engineering

The project engineer's role as described in Section C is more office than field based, but that office may also be at the jobsite. Although early design review is customarily the responsibility of more advanced team members, the PE will support estimators, schedulers, and project managers in a variety of fashions. Some of the activities described in Chapter 10 include value engineering, shop drawing review, and request for information (RFI) management. Chapter 11 continues with other document control activities performed by the PE, including submittals, assisting the superintendent with daily job diaries, meeting notes, and record document updates.

Chapter 12 builds upon many of the cost and schedule activities performed by the PE as introduced in Chapter 8, but now with a project manager and home-office support focus. Discussion in this chapter revolves around integration of the four pillars of project success; they must all work together to achieve a successful construction project. Earned value management is an advanced technique used to determine the cost and schedule status of an activity or the entire project. The PM develops a monthly cost forecast that the PE can assist with, as well as other home-office reporting activities.

Sustainability concepts are frequently implemented in BE projects today. Although it used to be a choice of some owners to implement some sustainability concepts, today many of these concepts are mandatorily addressed in the building code and others just make economic and moral sense. Sustainability is the topic for Chapter 13.

Chapter 14 includes a detailed discussion of project close-out. Close-out is often the role of the project engineer and the foreman because the PM and the superintendent have both moved off to another project. There are many close-out activities required before the contractor can receive its retention. The PE will participate in early close-out planning, commissioning, and construction close-out, including the punch list, contractual and financial close-out, and warranty management. Chapter 15 is dedicated to the topic of modeling project documents, including the use of BIM and civil information modeling (CIM). Chapter 16 focuses on updating estimates and schedules with current data.

Section D – Advanced project engineering

Many of the topics introduced in Section D of the book could be the subject of standalone books, and in most cases they are. Here, we simply provide a brief introduction to a set of advanced topics that senior project engineers will rely upon. In many cases, senior project engineers will actually function as assistant project managers; sometimes, they will even have the opportunity to run a smaller project. Most construction management books start with the basic estimating and scheduling foundation, but the entry-level project engineer will not have much of an introduction to these topics in the field until they have had a few years of experience. We therefore placed them in the last section of this book. As stated earlier, each chapter and each section of the book builds on its predecessors with increasing levels of responsibility for the PE. Although the new project engineer will not take the lead on many of the activities in this section, they will have opportunities to assist project managers, estimators, schedulers, and superintendents in the tasks and processes.

In Chapter 17, we differentiate between types of estimates: budgets, detailed estimates (including lump sum bids and unit price bids), and guaranteed maximum prices (GMP). We then walk the reader through the estimating process from developing a work breakdown structure (WBS) and performing quantity take-offs (QTO) through pricing and estimating jobsite indirect costs. We introduce schedule types in Chapter 18, differentiate between top-down and collaborative scheduling processes (including pull planning), and recommend procedures to obtain subcontractor and foremen buy-in to the schedule.

Most of the work on any jobsite is performed by subcontractors, so awarding subcontractor agreements and purchase orders to suppliers is an important first buyout step in the cost control process. Chapter 19 discusses the project engineer's role in subcontract development and the administration and jobsite management of these important second-tier team members. During construction, change orders (COs) will arise for a variety of reasons, and the management of the change order process is critical to the GC's success. In Chapter 20, we differentiate between change order proposals (COPs) and formal contract change orders and we describe the project engineer's role with gathering COP backup and assembling fair pricing.

There are a multitude of risks in construction management, and the project team's ability to understand sources of risks and cultivate methods to avoid them and transfer risks to other parties is an advanced project engineering and project management role. Chapter 21 introduces the critical topic of risk management. Our final chapter, Chapter 22, is both a summary of this book and an introduction to project management. Once project engineers have got this far, they are ready for the next step in their career path, which is accompanied by increased responsibilities. Many of the topics we introduced earlier in the book are included in project management books, but our focus throughout has been on the role of the project engineer. In this final chapter, we briefly expand those topics to the project management level and introduce a few advanced topics, such as claims, dispute resolution, and construction leadership.

Case studies

We have described two different case studies in Appendix A. The Rose mixed-use project is an apartment building in Portland, OR. This project is a wood-framed building over a cast-in-place concrete podium. The owner selected the GC on a negotiated basis and entered into a guaranteed maximum price contract with them. The Interstate 90 highway overpass project is a heavy-civil unit-price bid project just outside of Missoula, MT. Example documents from both of these projects are used throughout the book and additional documents are available on the companion website. Many of our review questions and applied advanced exercises also use information from these projects. In addition, we have included appendices for the abbreviations and glossary and an index to assist with navigation through an *Introduction to Construction Project Engineering*.

Summary

Construction project engineers work with other professionals toward the development of the built environment in the form of facilities and infrastructures. AEC industries are those operating toward this goal. Although their economic output fluctuates over time, it is often above 4% of the domestic GDP. Moreover, AEC industries employ above 5% of the domestic workforce. They employ professionals, such as architects and engineers, as well as craftsmen, such as electricians, ironworkers, and carpenters. Most of these careers are highly specialized, so their mastery should be recognized by correctly identifying and distinguishing one specialty from another. While facilities and infrastructures are the products of AEC industries, they are achieved through specialized processes shaped around projects. Projects vary largely depending on their intended products, but they can be classified into four major groups by product types: residential buildings, commercial and governmental buildings, heavy-civil facilities or infrastructures, and industrial facilities.

To provide students with a comprehensive overview of the duties and responsibilities of construction project engineers, this book is organized into four major sections. The organization of this book, from basic to more advanced topics, is, we feel, unique but logical for an introduction to construction management with a primary focus on the project engineer. The reader may have already noticed the abbreviations and technical terms introduced in this chapter. This use is common in construction, so we include an appendix for abbreviations. Additionally, we provide a glossary for most of the terms introduced in the book narration.

Review questions

1. Other than the examples we listed in this chapter, what are three additional examples of the built environment that would fall under the book's definition? What would be three examples that would fall under a broader definition of built environment?
2. List another specialty design firm that might work for a commercial project, beyond those listed in this chapter.
3. When might a heavy-civil project have an architect serving as a sub-consultant?

4.   Carpenters and plumbers have hammers and wrenches in their toolboxes. What sorts of tools do project managers and project engineers use?
5.   Why do you suppose new college graduates do not start their careers on day one as PMs?

References

NAICS (2017, May 10). North American Industry Classification System (NAICS) Main Page. Retrieved September 26, 2017, from www.census.gov/eos/www/naics/

BLS (2017, March 31). May 2016 National Occupational Employment and Wage Estimates. Retrieved September 26, 2017, from www.bls.gov/oes/current/oes_nat.htm

(2015, December 17). Construction and Extraction Occupations. Retrieved September 26, 2017, from www.bls.gov/ooh/construction-and-extraction/home.htm

2 Built environment projects

Introduction

The built environment (BE) industries produce and modify facilities and physical infrastructures through a broad set of activities aimed at adding to or changing the residential, commercial, governmental, industrial, transportation, and utility functions of the underlying natural, economic, and social environments. However, the underlying environments can make these tasks easy or hard, and each undertaking is uniquely different. As a result, construction projects are the standard approach to produce, renew, or modify facilities and infrastructures.

Activities can be project activities or mass-production activities. Projects are used by many industries as a way to organize activities to achieve a particular aim. Examples of projects include the development of a new medical device, the delivery of a healthcare program to address the specific needs of a target group, the attempt by a large company to set up a business branch in a new country, and the construction of a new apartment building. A popular definition of a project explains why certain activities, including most of those used to develop the built environment, revolve around projects. As defined by the Project Management Institute (PMI), a project is "a *temporary* endeavor undertaken to create a *unique* product, service, or result" (PMI 2016; p. 8). Mass-production activities are not projects because they are not creating a unique product. Moreover, mass production is not temporary because its duration is undetermined and depends on the marketability of a particular product, such as a specific model of an airplane. Therefore, activities leading to mass-produced goods belong to repetitive manufacturing processes.

The adjectives "unique" and "temporary" qualify a group of activities as a project. They also explain how project activities differ from the repetitive activities needed for mass production. Whereas various industries may rely on both project and mass-production activities, built environment industries mainly rely on project activities for the construction of facilities and infrastructures. A comparison against the aviation industry will explain the focus on projects among built environment industries. The aviation industry uses projects to develop working prototypes of airplanes for mass production. In aviation, a prototype would be the first instance of an airplane, so this prototype is *unique*. Moreover, the group of activities necessary to develop the prototype is *temporary* because the project ends once it reaches its objective. Afterward, the mass production of airplanes based on the prototype may result in many thousands of instances of the same airplane. By contrast, a typical construction project aims for the delivery of a facility or infrastructure and ends once this facility or infrastructure is completed. However, mass-produced goods, such as electrical outlets and door handles, are also necessary to produce the built environment. Some BE projects also follow repetitive processes similar to those in manufacturing, as in the case of large-scale single-family home subdivision projects.

Nonetheless, most construction projects are largely different from manufacturing. As built environment facilities and infrastructures are the intended final product and will not be mass-produced, the activities to create them are *temporary*, which means that they are assigned a defined scope, a clear timeframe, and a set of resources, such as people, equipment, and materials. Moreover, each facility or infrastructure project is *unique* to its underlying natural, economic, and social environments.

The built environment and its underlying social, economic, and natural environments are related. The term *sustainable project* is used to define the ability of the facility or infrastructure to strike a balanced fit with its underlying social, economic, and natural environments. This topic is becoming more important to construction and project engineering tasks and will be discussed in detail as part of Chapter 13.

Example

A road project may be initiated by a city to address an increase in traffic due to population growth in a suburban area. This project will need to produce an infrastructure that fits within its underlying social, natural, and economic environments:

- *Social environment*: Studies will be performed to evaluate how the new road will benefit the public, and outreach activities will be conducted to involve and/or inform the public during the design and construction process;
- *Natural environment*: Environmental studies will be performed to evaluate if the impact on the natural environment is overriding the expected advantages of the new road. These studies will also suggest mitigation measures that would be necessary to reduce or nullify these impacts during construction; and
- *Economic environment*: The impact of the project on the local economy will be evaluated, including the impact of the cost of this project on the city economy at completion. Also, the impact of construction activities on existing businesses will be evaluated, and mitigation measures may be needed during construction to minimize this impact.

Another aspect of a construction project's uniqueness is the multitude of organizations and people working together to achieve the project objectives, namely, the project team that changes from project to project. In terms of organization, facilities and infrastructures are usually developed by *project owners* for their *end users*, designed by *design firms*, and built by *construction firms*; these are the four main participants in a BE project, as will be discussed in Chapter 3. For instance, a multi-billion dollar new highway turnpike system around a state capital may be needed to relieve city traffic while providing a way around the city for pass-through vehicles. This system will be developed by the state transportation agency on behalf of the state's citizens. The turnpike would probably be delivered through many road projects, each staffed by a different set of designers and builders.

Construction projects are sometimes related to each other because some project owners may decide to fulfill a larger set of needs through a series of projects. For instance, due to the magnitude of the previously cited turnpike, it may be difficult for the state transportation agency to identify design and construction firms that are large enough to handle all of the risks associated with such an undertaking. Instead, the turnpike could be delivered through a series of separate projects, which will need to be well coordinated. Breaking down a large undertaking into smaller projects is common for infrastructures or groups of facilities. When this approach is used, it is common to define the entire undertaking as a *program* or *masterplan*, which is to its construction projects as an entire puzzle is to its pieces. A program would provide its BE projects with the general guidelines necessary to achieve

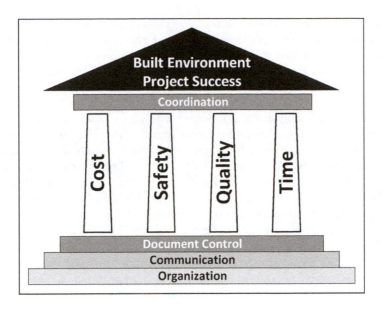

Figure 2.1 Evaluating Project Success

the overarching set of needs while leaving each project to define its own set of objectives that does not conflict with these guidelines. For simplicity, we will only refer our discussion to individual construction projects as part of this book.

Evaluating project success

Evaluation of the success of BE projects is sometimes a challenging task for numerous reasons. First, projects are usually delivered through efforts by multiple parties, each having different priorities and objectives. Using a generic project as an example, we can discuss these conflicting objectives and priorities. The project owner of a commercial development will consider a project successful if it achieves its objectives, is delivered on time, and does not cost more than planned. The design team will consider the same project successful if it provides the design firm with a monetary reward and results in a quality facility that will enhance their reputation. Finally, the construction team will consider the project successful if it provides a monetary reward and does not result in injuries or casualties. Therefore, different parties may come to work on a project with different priorities. Any BE project is initiated by a project owner who should set the project objectives, but the project is developed by other project parties on behalf of this project owner. Achievement of the owner's objectives will accomplish overall success of the project, but it is the owner's responsibility to convey their objectives to the other project parties and to facilitate their alignment. Therefore, these project objectives should be clearly identified as soon as possible at the project onset.

Second, project objectives may differ greatly across construction projects. Yet it is still important that the project parties speak the same language when it comes to evaluating project success. To this end, the success of built environment projects is traditionally evaluated along four measurable dimensions that constitute the pillars for success of BE projects in the United States: *cost*, *time*, *quality*, and *safety*, as shown in Figure 2.1. *Cost* relates to achieving the project objectives within the provided budget. *Time* relates to achieving the objectives within the provided timeframe. *Quality* relates to meeting the project owner's expectations and needs, as well as all existing codes. *Safety* relates to achieving the objectives without accidents or fatalities.

However, true project success can only be achieved through a balanced performance of all four of these dimensions. For instance, it would be difficult to define a project as successful if anyone died during construction, even if the final facility was completed on time and on schedule and met the quality expectations. Similarly, a significant cost overrun would make it difficult to claim success for a project that was completed on time, up to the quality expectations, and without safety issues. However, all parties may assign different importance to each of these dimensions, which may make it difficult to maintain the desired balance throughout all of the project phases. Still, a coordinated effort to translate priorities by individual parties into measurable benchmarks along each of these dimensions is crucial to project success. This ability to *coordinate* parties acts as a binding agent and is often a precursor to project success.

Finally, not everything is measurable, and several additional factors act as the foundation to project success, such as the ability to *organize the project team* by clearly identifying the roles and responsibilities of each party, the ability to establish and maintain successful *communications* among project parties, and the ability to record communications through appropriate *document control*.

Project phases

As previously stated, facilities and infrastructure projects are usually developed on behalf of *project owners*, who are sometimes also the *end users*, designed by *design firms*, and built by *construction firms*; these are the four main participants in a BE project. Once a project owner has identified that a built environment undertaking will help fulfill their needs, a project is initiated and a process will take place that will result in the completion of the facility or infrastructure and its placement into operations.

Since many built environment projects require significant time and financial investments, they are usually phased. Moreover, many owners follow a phase-gate model to authorize expenditures on projects, where each project has a gate at the end of each phase. The decision on opening the gate to the next phase is usually based on what was produced in the previous phase. If the gate is opened, funds and time will be provided for carrying out the following phase.

In the remainder of this section, we will use examples from the private and public building sectors to define and describe the main phases of a built environment project. Although the terminology for infrastructure and industrial facility projects may differ, these projects are delivered through a similar sequence of project phases, which is shown in Figure 2.2. These major phases include planning, design, construction, and operations.

Planning

During this initial phase, the seed for a potential new built environment project is planted (i.e., project initiation) and a project concept is developed (i.e., conceptualization), as shown in Figures 2.3A and 2.3B. This phase occurs once a project owner realizes some needs or wants, evaluates alternative approaches to achieve these needs, including approaches that would require the initiation of a new built environment project, and generates a project concept. Depending on the industry sector and the project owner, other known names for this phase are *programming*, *pre-design*, *front-end planning*, and *pre-project planning*. This phase is performed by the owner and should not be confused with the pre-construction planning and start-up phase that is performed by a general contractor.

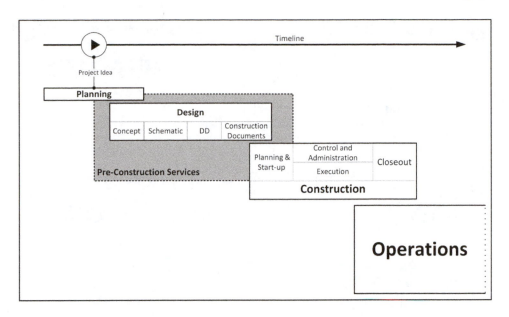

Figure 2.2 Major Phases of a BE Project

In the private sector, this phase is part of the project owner's *business planning* process. For instance, while performing its business planning, a private industrial company may realize the need to increase production of its goods to fulfill an increase in market demand in a remote region. The company may increase production at an existing industrial facility by increasing work shifts or build a new facility to fulfill the increase in demand. To compare these options, the company needs to develop a rough concept of the new facility to be used to estimate its costs. After performing a comparative analysis of these two options, the company concludes that initiation of a project to build a new factory in proximity to the target market would be preferable and more remunerative than an increase in shifts at the existing industrial facility.

In the public sector, this phase is part of the *community* or *regional planning* process. For instance, while performing its community planning process, a city may realize the need to reduce its fire department response time in a certain neighborhood. The city may add fire trucks at a fire station in a nearby neighborhood or build a new fire station in the given neighborhood. To compare these options, the city needs to develop a rough concept of the new fire station to be used to estimate its costs. After performing a comparative analysis of these two options, the city concludes that initiation of a project to build a new fire station would be preferable, because the addition of fire trucks to the existing station would not provide adequate response times during rush hours.

As described in these examples, during the project planning phase, alternative approaches to fulfill needs are generated, analyzed, and compared. The alternatives may envision and conceptualize new built environment undertakings. If a new built environment undertaking is selected, the project is initiated and based on the concept used to make the decision. As part of this phase, different facility sites or infrastructure alignments may be investigated as alternatives. The project planning phase is typically led by the project owner with their own personnel, even though consultants may be hired to carry out some analyses and support the owner's decision-making process. One of the

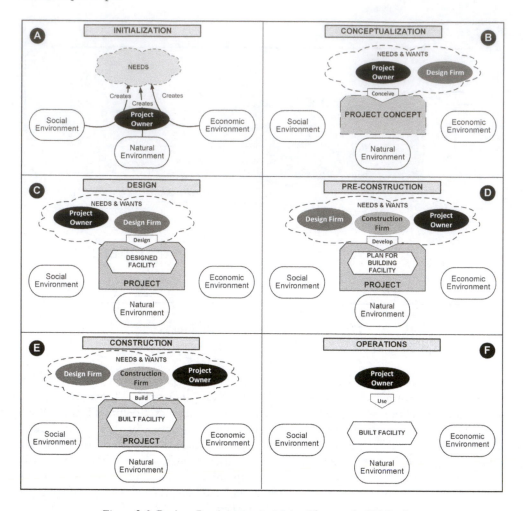

Figure 2.3 Project Participants in Major Phases of a BE Project

most common types of analysis performed at this stage is the *feasibility analysis* (or Study), which evaluates the proposed project's ability to fulfill the needs, as well as the technical and economic feasibility of the project concept.

Design

The design phase, shown in Figure 2.3C, can be divided into four subphases, namely, *conceptual design*, *schematic design*, *design development*, and *construction documents*. They are often part of the phase-gate model used to authorize expenditures.

An initial *conceptual design* (i.e., *project concept*) is actually developed during the planning phase as part of the conceptualization of the project idea. Conceptual design may be led by the project owner with their own personnel, but a design consultant is usually hired to help develop a concept based on sound design principles. Based on the project concept, a preliminary cost estimate is usually performed at this time.

Once a project idea is formed around a project concept, the next step is to develop a *schematic design* that expands the project scope, which can be used as the basis for later design development. It is common for project owners to assemble an early design team that will develop a set of graphical documents that add detail to the initial concept. As part of this subphase, it should be verified that the facility concept can be built on the available site, and if not, the conceptual design should be revised. At the end of this phase, the owner will be provided with a schematic design package and a cost estimate for review and approval. Schematic design usually results in a designed facility with dimensions only at the macro level. Although it is becoming more frequent for designers to develop three-dimensional building models by using computer-aided design (CAD) or building information modeling (BIM) software, the design is usually provided to the owner in the form of drawings, including a site plan, floor plans, sections and elevations, and printed renderings. Moreover, the design team often provides a presentation to the owner where three-dimensional information and simulated walkthroughs can be shown. Chapter 15 includes information on three-dimensional information modeling techniques. The cost estimate is based on the most reliable level of dimensional information provided in the schematic design package and is usually based on estimated floor areas or building volumes.

Once the schematic design has been approved by the owner, the design firm will proceed with the *design development* (DD) subphase. As part of this phase, the design firm will work toward the production of a design package that will outline the specifications and include architectural information, such as floor plans, sections, and elevations, as well as layouts of structural, mechanical, electrical, and plumbing systems. These drawings will be produced at a higher scale than those with the schematic design. They will include full dimensions and provide door and window details, as well as information on materials. The design package will be submitted for review and approval to the owner. Often, the design firm would also provide a presentation to the owner in which additional content can be provided, including updated cost estimates, a preliminary construction schedule, three-dimensional renderings, and results of analysis on the owner's "hot button" issues.

Once the owner and design firm complete the review and revision of the *design development* package, the design team proceeds with the final stage of design to produce the *construction documents*. Construction documents take the design to a greater level of detail to include construction details and specifications for materials; they are used to finalize the price of construction. The level of detail achieved in the contractual document set is decided by the owner and may vary greatly with different project types and project delivery methods.

Construction

Project owners rely on the construction team to lead the construction phase. During this phase, the construction team members first plan and prepare for the construction process. Later, they perform and/or oversee field activities and carry the responsibility for the outcome. Traditionally this phase can be subdivided into four subphases:

* Construction planning and start-up (also known as *traditional pre-construction*),
* Construction execution,
* Construction control and administration, and
* Construction close-out.

The main goal of the *construction planning* and *start-up* subphase is to plan and prepare for construction, as shown in Figure 2.3D. On projects delivered with a design–bid–build delivery (see Chapter 4), this subphase also incorporates some pre-construction activities as follows:

- pricing,
- project scheduling, including logistics and phasing,
- pre-qualification of subcontractors and suppliers,
- solicitation and selection of specialty contractors,
- analysis and purchasing of materials and products,
- review of construction documents, including constructability analysis, and
- assistance with obtaining sustainability certification.

Recently, the term *pre-construction services* has grown in popularity to identify a set of activities broader than simple construction planning and start-up. These additional activities are allocated to the construction team when it is brought on board early on a project. Through this early involvement, the team is expected to improve the plan for construction. Therefore, activities traditionally performed as part of construction planning and start-up are included as part of a larger à-la-carte menu of services that project owners can elect to purchase from the construction team. Under these circumstances, activities that were traditionally self-performed by the project owner staff or by the design team will be totally or partially performed by the construction firm. Examples of pre-construction services provided during various phases may include:

- Planning (the contractor is not involved in this phase except in rare circumstances and project types):
  - scope definition,
  - selection of design team and responsibility to coordinate design and construction team members, and
  - selection of project delivery method.
- Design:
  - site analysis and evaluation,
  - investigation into local codes and zoning rules,
  - value engineering analyses,
  - constructability analysis at all stages of design and for all systems,
  - design phasing and coordination,
  - early procurement,
  - safety planning,
  - quality management planning,
  - life-cycle cost analyses, and
  - scheduling, estimating, and pricing at all stages of design.

The *construction execution* phase will deliver the facility or infrastructure as it was designed, as shown in Figure 2.3E. This phase is focused on field activities that will be discussed in Section B of this book.

The main goal of the *construction control and administration* subphase is to oversee the construction execution and to enforce coordination among the project parties. This phase is focused on jobsite office activities that will be discussed in Sections C and D of this book.

The *construction close-out* subphase will complete all of the construction tasks and gather all documentation required to close out the construction contract and complete the project. This process may take up to a year after the certificate of occupancy is issued and incorporates four concurrent processes: field close-out, contractual close-out, financial close-out, and in-house close-out analysis. This phase and its subphases will be described in detail in Chapter 14.

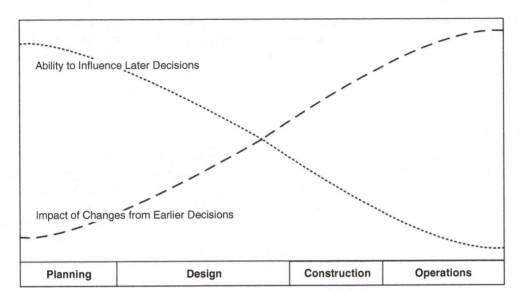

Figure 2.4 Influence Curve

Operations

Once construction is complete, the facility or infrastructure is ready to be put into operations, which means the project owner can start using it, as shown in Figure 2.3F. For many projects and most of the participants, the project objectives are fulfilled at construction close-out, so the operations phase is not part of the project, although operations is the next phase of any facility or infrastructure life cycle. However, a facility or infrastructure may be the object of several other projects during its lifetime, after its initial execution. Typical post-completion project objectives include retrofitting to add functions that were not provided in the initial project, performing extraordinary maintenance to the whole facility or infrastructure, revamping to improve the underlying design, or demolishing it to restore the initial natural conditions or to prepare the land for other uses.

Impact of early decisions on later expenditures

Among the many slogans regarding how to achieve project success, one standout for built environment projects is to "do it right the first time." This message captures an important truth about BE projects: decisions made early on have a greater impact on the project outcome over the facility or infrastructure lifetime than later decisions. This concept is graphically illustrated by the influence curve shown in Figure 2.4 and is explained through the following examples.

Planning: Infrastructure alignment or facility location

One of the main decisions to be made during the planning phase relates to where the built environment facility or infrastructure will be located. For example, a road alignment would be identified at this stage and used to perform all the impact analyses required for a permit. Similarly, the location of a new facility will be decided at this time. Any error in this phase will affect the ability of the facility or infrastructure to fulfill its needs. Moreover, any later change to decisions made at this stage may have substantial to catastrophic impact on the project. A change in a road alignment may require new studies to be performed and an amendment to the initial permit to be obtained. Similarly, the later realization that a project site may not be fit for its intended use will force the project owner to identify and acquire a new piece of land, obtain new permits, and redesign the facility to the new site constraints.

Early design: Building layout

Some of the major decisions to be made during the early design phase of a new building are those related to its siting, including its orientation and layout. Decisions made at this stage heavily affect all of the subsequent phases. Incorrect decisions during this phase may affect the ability of the facility or infrastructure to fulfill its needs, but any change to these early decisions would have a significant impact on the project's cost and schedule. For instance, a realization that the building has been designed with the wrong orientation or layout will require either acceptance of a faulty design for construction or a redesign of the facility.

Late design: Detailed design of bathrooms

This is probably the latest stage when decisions with an impact on the project may be changed. For instance, at this time, the bathroom layout will be finalized, along with the layout of its mechanical, electrical, and plumbing (MEP) systems. Since public bathrooms need to be compliant with the Americans with Disabilities Act (ADA), any error may affect the project's ability to meet ADA code requirements. Still, this is the time to implement changes to the design without construction rework or constraints on the building's use.

Pre-construction: Selection of specialty contractors

One of the major decisions to be made at this stage is the selection of specialty contractors for various items of work. Whereas these decisions do not have catastrophic effects on the usability of the building or infrastructure, they may hit the project's cost, quality, and time performance. For instance, selection of a painting contractor without the necessary manpower to deliver the scope may result in quality issues and delays. This would result in potential claims for additional compensation or penalties for late completion of the project.

Construction: Changed delivery schedule of materials

Ideally, all major decisions on the construction of a building should be made during the pre-construction phase. However, some decisions made during the construction phase may create significant issues. For instance, on a high-rise project, the delivery of materials needs to meet the requirements stated in the traffic control permit and is often assisted by the project's tower crane. Therefore, if a delivery needs to be changed, the permit may need to be modified and the new delivery needs to be verified against the availability of a time slot for the tower crane.

Project engineering applications

As will be described in Chapter 3, project engineers work for a construction firm. Therefore, they will be involved in many of the project phases when a construction firm is onboard. Project engineers include entry-level employees and professionals with several years of experience. Most of the less-experienced project engineers will be involved in the central phases of construction, including project execution, control, and administration. Participation in these phases will provide the necessary knowledge of the project and experience to allow the same project engineers to assist the project manager in the close-out. More experienced project engineers will also be beneficial in pre-construction, when they are often delegated specific sections of the project planning and start-up. When pre-construction services are associated with early contractor involvement, highly experienced project engineers can be called upon to assist the project manager in providing early pre-construction services, such as estimating, scheduling, and procurement, based on their knowledge and experience.

Summary

Facilities and infrastructures are developed as *projects* due to the *uniqueness* of the underlying natural, social, and economic environments and the *temporary* nature of the undertakings. Once initiated, BE projects require multiple parties to *coordinate* the achievement of *project success* that ultimately accomplishes the project objectives, which are project owner-dependent. However, built environment industries have developed a common language to allow all parties to align and therefore refer to the four measurable pillars of project success as *cost*, *time*, *quality*, and *safety*.

Similarly, all BE projects follow a sequential phasing of activities and undergo four major phases, namely, *planning*, *design*, *construction*, and *operations*. Project engineers operate on a project as employees of construction firms. Experienced project engineers support the project management team in the traditional pre-construction phase of planning and for actual construction. However, recently, project owners are trying to secure early contractor involvement in the form of additional pre-construction services. Thus, pre-construction often expands broadly and becomes a new phase of the project that overlaps with design and, rarely, planning. Since early decisions can have immense impact on the project success, only highly experienced project engineers will be called upon to support the project management team in these broader activities.

Review questions

1. What are two characteristics that define a project?
2. What are the four typical design phases of a construction project?
3. What are the four dimensions to measure project success?
4. What other factors act as the foundation to project success?
5. Provide three examples of pre-construction activities during planning.
6. Provide three examples of pre-construction activities during design.
7. List the construction subphases of a project.
8. When do less experienced project engineers get involved on a construction project?

Exercises

1. Provide examples of issues that may affect project success in any of the four dimensions.
2. Perform an Internet search on the following construction issues:
   a. *Safety and accidents*: For instance, you could use the search string "construction accident" on Google News or Bing News. Select an article on a project and read it. Submit a short summary of what happened and on what type of project.
   b. *Cost overruns*: For instance, you could use the search string "construction cost overruns" on Google News or Bing News. Select an article on a project and read it. Submit a short summary of what happened and on what type of project.

c. *Delays*: For instance, you could use the search string "construction delay" on Google News or Bing News. Select an article on a project and read it. Submit a short summary of what happened and on what type of project.

d. *Quality issues*: For instance, you could use the search string "construction defects" on Google News or Bing News. Select an article on a project and read it. Submit a short summary of what happened and on what type of project.

Reference

PMI (2016). PMI Lexicon of Project Management Terms, Ver. 3.1. Retrieved September 26, 2017, from the Project Management Institute (PMI) website: www.pmi.org

3   Participants in built environment projects

Introduction

As discussed in the previous chapter, all built environment (BE) projects follow a similar life cycle. In this chapter, we learn that these projects also involve the same types of participants, which include businesses, organizations, and individuals. Most of the project participants are involved in multiple projects. However, like actors in movies, project participants can serve different roles on different projects. This chapter describes different ways to classify project participants, as well as some of the more common roles present on BE projects. A relationship between the organizations and individuals exists and is easy to capture by considering that organizations employ individuals, including construction project engineers. Moreover, organizations may be liable for the actions of their employees, and their legal status affects the exposure to liability for business owners.

In terms of classifications, project participants can be defined by their *functional role*, *legal status*, and *contractual role*. In this chapter, we will first discuss organizations operating in the built environment based on their functional role. Then, we will discuss businesses by their legal status. Last, we will discuss the relationship between organizations and individuals. The contractual roles will be discussed in Chapters 4 and 5.

Organizations

Each project will have a different team of organizations that will operate under a web of contractual relationships, also known as contractual frameworks. Chapter 4 will provide insight on different approaches for weaving a contractual framework. Chapters 5 and 19 will describe how organizations are selected for the project team and how contractual agreements are used to bring them into a project.

Construction projects are usually developed on behalf of *project owners* and *end users*, designed by *design firms*, and built by *construction firms*; these are the four main functional roles on any BE project. Other organizations include *insurance companies*, *sureties*, *authorities having jurisdiction* (AHJs), and various consulting and technician firms, such as *land surveyors*. The remainder of this section will describe the functional role of these organizations on a built environment project.

Project ownership and use

Each project has its own *project owner* who evaluates the *end user's* needs before establishing the project scope and objectives, sets a timeline to achieve these objectives, and secures funds for the project. Oftentimes, a *project owner* is also the *end user*.

In this book, the term *client* is sometimes used in referring to the project owner. Although in each project contractual relationship, we can identify a client and a provider, the term client will

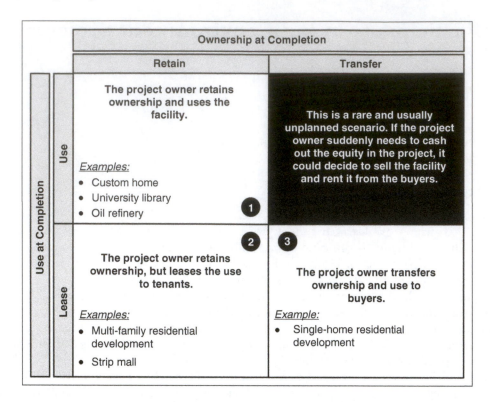

Figure 3.1 Ownership and Use at Completion

be used in this book only to refer to the project owner in the context of direct relationships with design consultants and contractors. The concept of ownership may also apply to owners of companies. For simplicity, we will use the shorthand term *owner* or *client* when referring to the project owner for the rest of this book, except when it is clearly stated otherwise.

An important difference among project owners is their funding source—public versus private. A homebuilder is an example of a project owner who fully finances their projects with private funds. A city is an example of a project owner that fully or partially finances its projects with public money. Project owners who rely on public money are called public owners and are subject to strict rules to allow for transparency and fairness. As shown in Figure 3.1, two characteristics affect a project owner's functional role on a project: (1) the owner's plan for using the facility or infrastructure at project completion, and (2) the owner's plan for retaining ownership at project completion. These characteristics correspond to the three main scenarios shown in Figure 3.1, which include build-to-use, build-to-lease, and build-to-sell projects. Some examples are provided to clarify how these characteristics may manifest and affect a project.

Scenario 1: Build-to-use projects

Some project owners build facilities for their own use. Therefore, they will not only retain ownership at completion, but they will also be *end users* of the facility. Private citizens, governmental agencies, and companies that need facilities to support their core business often build these facilities for their own use.

Examples

- A prospective homeowner buys a piece of land and has his or her dream home built on it.
- A university needs a new library building and decides to have it built on a vacant campus lot.
- An oil company needs to revamp an existing refinery that it has just purchased.

As the end users of facilities, owners should be able to identify the needs that the project must fulfill and set the project objectives accordingly. Moreover, they will be motivated to build something that will reduce costs during the operations of the built facility. Therefore, the homeowner will probably attempt to obtain a durable and comfortable home, the university will seek a library building that is both durable and inexpensive to operate and maintain, and the oil company will seek a facility that is easy and safe to operate while limiting the need for extraordinary maintenance.

Scenario 2: Build-to-lease projects

Some owners initiate projects to build facilities that they will own at completion but will lease to short- or long-term tenants. Build-to-lease projects are usually those carried out by real-estate developers whose main core business is developing and managing real-estate properties. The decision to retain ownership at completion is sometimes only temporary because some developers decide to build and lease for a few initial years of the facility's life cycle but then sell later. The lease terms will change greatly depending on the facility type and will state the developer's responsibilities in terms of operations or maintenance costs. Depending on the facility type and market, tenant expectations may vary quickly over time and among tenants, which will ultimately affect the project objectives.

A developer of a build-to-lease project may expect a high tenant turnover and large variation in expectations from tenant to tenant for the facility, which will motivate them to build a facility that can be easily adapted to the changing needs of commercial tenants. Under these circumstances, the facility will not be completed until a tenant for a retail or office space is selected, which results in phases of construction. The first phase will be to design and build a *core and shell*, including the building structure, envelope, site work, and building systems. Once a tenant is secured for the space, a *tenant improvement* (TI) phase will be carried out that will focus on finishes, partitions, and trimming of mechanical, electrical, and plumbing (MEP) systems. The TI phase can be performed by the owner on behalf of the tenant or by the tenant. For facilities operating in the long-term lease market, developers will attempt to secure tenants before the core and shell phase is completed, which may result in late changes to the project scope to accommodate specific needs by early tenants that the owner has secured.

Examples

- A land-use commercial developer and property manager develops a new strip mall in a suburban area to be subdivided into commercial units for lease. This owner is motivated to develop a building that is durable and fits the needs of the prospective tenants. Since the facility life is expected to be much longer than the average length of a lease, tenants are expected to change over time. These characteristics suggest a phased approach, with the core and shell separate from the TI.

> • A similar situation would be represented by a residential property developer committed to building and owning multi-family properties. This owner is also motivated to develop a building that is durable and fits the needs of the prospective tenants. The building should also be easy and economical to maintain and operate because the developer will perform maintenance on the apartments and common areas, as well as operate them. However, this project would not need to be phased because apartment finishes are identified by the target tenant market (e.g., luxury, mid-range, or short stay), which is often dictated by the location of the building.

Scenario 3: Build-to-sell projects

Some owners build facilities to sell them at construction completion. Build-to-sell projects are usually those carried out by real-estate developers whose main core business is to buy and develop land and then profit from the increased value. Since these developers will not be involved in the operation and maintenance of the built facility, lifetime considerations for durability and operations and maintenance (O&M) costs may be limited to their target client's interests.

> *Example*
>
> • A homebuilder acquires a plot of land, applies for permits, and develops a new single-home suburban subdivision while selling 200 individual homes to prospective homeowners and constituting a homeowner association to manage the common areas.

Design firms

Project owners rarely design their own facilities. Most projects are designed with the support of *design firms* that will devise a facility to meet the end user's needs and the owner's time and cost constraints. Simple projects may be designed by a single design firm, but design expertise is highly specialized, so it is common for a project to require services by more than one design specialty. Design firms are businesses that employ licensed design professionals. Licensures are specific to design specialties, such as architecture, electrical, mechanical, or civil engineering. Moreover, licensure is state-specific, so a building project based in the State of Michigan will need an architect licensed to operate in that state. Larger design firms will employ licensed professionals in many disciplines, but smaller firms may only have a few employees.

In the United States, it is customary for an architecture firm to lead the design process for building projects with the support of consultants from other design specialties. The design activities for civil projects are usually led by a civil engineering firm, which will secure consultants in other specialties as needed to design the project scope. Similarly, the design activities for industrial projects are usually led by industrial engineering firms that will design the industrial processes and secure consultants in other disciplines as needed to build a facility that will host these processes.

Construction firms

Project owners rarely build their own facilities. Instead, most projects are built with the support of *construction firms*. Whereas these firms vary largely depending on the industry sector in which they operate, most of them would fall into one of several major groups.

- *General contractors (GC)*: These construction firms are responsible for the overall planning and coordination of construction operations. They are generalist firms that have knowledge of the whole process to provide overall supervision and coordination. They usually perform some work themselves and hire specialty contractors for other work items. General contractors employ jobsite office and field management personnel, as well as a specialized workforce in the scopes they self-perform. The amount and type of scope a GC would self-perform is highly dependent on the project type and the contractor, and there are some general contractors that do not self-perform any construction activity. For instance, a GC operating in the commercial building sector would frequently self-perform structural work but would hire electrical, mechanical, and plumbing specialty contractors. Therefore, this GC will employ project managers, project engineers, superintendents, and foremen. In addition, they will employ craftsmen, such as carpenters and laborers. This firm may own some construction equipment but will most likely lease specialized and expensive pieces of equipment. However, if the same GC is hired to perform work in a location outside of its primary business area, it may decide to hire a local structural specialty contractor instead of employing new skilled workers.
- A *construction manager* is similar to a general contractor, except they do not self-perform scope and, therefore, do not hire direct labor.
- A *heavy-civil contractor* operating in the road construction sector would probably self-perform most direct tasks, including earthmoving and paving work. This firm will employ project managers, project engineers, superintendents, foremen, equipment operators, and laborers. They also often own a fleet of construction equipment to support their self-performed operations.
- *Specialty contractors*: These construction firms are responsible for specialty areas. These firms employ a highly specialized workforce and own specialized tools and equipment to support the tasks they perform. Examples of specialty contractors include earthmoving, electrical, mechanical, plumbing, and painting contractors.
- *Suppliers*: These firms provide only materials and equipment for the project. Their contribution is usually limited to supporting the installation by others of the materials or equipment they provide. When required to perform installation, they would act as subcontractors and would need to provide liability insurance.
- *Agency construction managers* coordinate the work of prime contractors who contract directly with the owner but do not perform work themselves.

Other participants

Many other entities and individuals participate in built environment projects, even if they do not directly contribute to the activities necessary to complete a project. This section provides an overview of a few selected participants that will frequently be encountered on construction projects.

An AHJ is defined in the National Electrical Code as "an organization, office, or individual responsible for enforcing the requirements of a code or standard, or for approving equipment, materials, an installation, or a procedure" (NFPA 2017; pp. 33–34). For instance, a city will have jurisdiction on construction activities within its boundaries. Therefore, a project located within the city's boundaries will need to obtain numerous permits before commencement. The project will need to abide by city codes. City officials will review the application for permits and issue the permits if the application is in compliance with the city codes. Once construction activities are in progress or completed, inspections are performed to evaluate if the work is compliant with the city codes. There are many AHJs, and the location of a project and its type will dictate which AHJ will apply to it. Other examples of AHJs are federal, state, and local agencies, counties, and Native American sovereign nations.

Insurance companies and *construction sureties* are companies that are willing to take some of the liability associated with the design and construction activities for a fee. These entities rarely intervene in the project and only in response to the occurrence of specific events. There are some important differences between insurances and sureties. The following paragraphs provide an overview of these differences, which will also be discussed within Chapter 21 on risk management.

- *Insurance companies* use analyses to evaluate the chances that an event may occur and its financial impact on a project and to establish a premium price for them taking on that risk. Risks are usually grouped by category and packaged into standard insurance company policies. An insurance policy is a contractual document between the insurance company, the insurer, and the entity or individual that requests coverage against the risks, the insured. Examples of standard insurance policies for the built environment include professional liability insurance, errors and omission insurance, general liability insurance, property insurance, builder's risk insurance, business owner insurance, worker compensation insurance, and construction vehicle insurance.
- *Construction sureties* investigate the project and its participants to evaluate if the chances of an event occurring are low enough to issue a bond. Three types of bonds are prevalent in the industry: bid, performance, and payment bonds. By issuing a *bid bond*, a surety is assuring the bid-receiving party (often the project owner) that the bid-submitting party (usually a construction firm) is submitting its price proposal in good faith and will enter into a contract for services at the proposed price and that the surety will provide performance and payment bonds if they are needed. By issuing a *performance bond*, a surety is assuring the service-receiving party of protection from economic losses in case the service-providing party (usually a construction firm) is not able or willing to fulfill the contractual expectations. By issuing a *payment bond*, the surety is assuring the service-receiving party that the service-providing party (usually a construction firm) will pay subcontractors, direct labor, and material suppliers on the project. Unlike insurance policies, which are agreements between two parties, bonds are agreements between three parties: the principal, the obligee, and the surety. Figure 3.2 describes the relationship between these three parties and provides an example of a bid bond.

Land surveyors are licensed professionals who perform construction surveys, to identify the position of the proposed facility or infrastructure, and other surveys, such as site utilities and topography mapping. Owners, design firms, and construction firms rely on the services of land surveyors throughout a project life cycle because it is necessary to establish property boundaries, terrain slopes, and building work positions.

Business legal status

Project participants mostly interact under contractual agreements, which are legal documents stating the duties and obligations of the parties. However, participants in projects can be human beings or business entities. Once they are created, businesses are given rights and responsibilities, some similar to those applicable to individuals. Different forms of businesses exist in built environment industries. The most common forms of business are sole proprietorships, partnerships, corporations, limited liability companies (LLCs), and joint ventures (JVs). Note that the concept of ownership is used in this section when referring to *business ownership*, in contrast to the rest of the book where ownership always refers to *project ownership*.

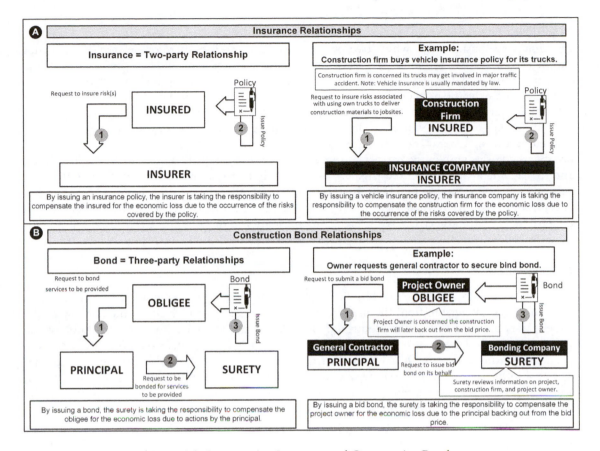

Figure 3.2 Construction Insurance and Construction Bonds

Sole proprietorships

Sole proprietorships are recognized as the most common type of business operating in the built environment industries. A sole proprietorship is popular because it is easy to start and maintain a business that is owned and managed by one individual, the sole proprietor. Therefore, there is no distinction between the business and its owner, who is entitled to all profits but is also fully responsible for the actions of the business. Requirements for forming a sole proprietorship change depending on the state jurisdiction; this action may not require any formal filing with the state. However, all permits and licenses required to operate in an industry or a location still need to be obtained by the proprietor. For instance, a handyman can operate as a small painting business, but he would still need to obtain insurance and a commercial license if so dictated by the jurisdiction where he operates. Additional advantages of operating as a sole proprietorship include its flexibility (because the business can go where the owner wants), its speed or response (because all decision-making is done by a single person), and its easiness to terminate (because the owner can decide at any time to terminate the business or sell it). The disadvantages of operating a sole proprietorship include the unlimited personal liability, the limitations in raising money to grow and/or operate the business, and the heavy decision-making burden, which falls on the sole proprietor.

Partnerships

Partnerships are businesses in which two or more people or entities share ownership. Partnerships can be seen as a shared form of proprietorship because partners contribute to the business in the form of resources and talents. Similarly to a proprietorship, a partnership is closely associated with its owners, namely, the partners. Also, profits are passed through to the partners and are taxed with other personal gains. However, a partnership can independently own property, hire employees, and undertake (or be the subject of) legal action. Partners can be one of two types: general and limited partners. General partners share the responsibilities, profits, and losses of the business. If not specified otherwise in the partnership agreement, partners equally distribute profits, liabilities, and management duties. Limited partners only contribute assets and do not participate in managing or operating the business. In return for their limited involvement, limited partners do not participate in liability beyond their initial contribution. The addition of limited partners is a way for a partnership to raise capital while providing immunity to the investor. Every partnership must list at least one general partner.

Corporations

Corporations are legal entities owned by shareholders. Forming a business as a corporation is a way to separate the ownership from the management and operations, which limits the liability to the stakeholders for only their initial contribution. Stakeholders appoint a board of directors, which selects and oversees business operations led by officers, including a chief executive officer (CEO), a chief operating officer (COO), and a chief financial officer (CFO). Unlike proprietorships and partnerships, corporations file taxes separately from their stakeholders. The profits of a corporation are taxed at a corporate tax rate, which is often lower than the personal income tax rate. If the management decides to distribute part of the profits among the shareholders, these distributions, or dividends, are subject to double taxation at the individual level. Two types of corporations exist: Type C and Type S. Type C is as described above, and Type S corporations are specially created through an internal revenue service (IRS) tax election to avoid double taxation on dividends.

Limited liability companies (LLC)

Limited liability companies combine features of corporations and partnerships. Owners of an LLC are referred to as members. Similarly to a corporation, an LLC would limit personal liability from ownership. Similarly to a partnership, an LLC would avoid double taxation by passing through all of its profits and losses to its members.

Joint ventures (JV)

Joint ventures are business entities that resemble general partnerships but can include in their partnership natural or juridical persons or governmental agencies. The main characteristic of a joint venture is its temporary scope because JVs are created for only a limited period of time or for a single project. The main reasons for using a JV on a project are to pool financial, physical, and human resources and for spreading risk. For example, many mega-infrastructural projects are built by joint ventures of construction companies that pool their personnel, equipment, and bonding capacities while spreading the risk of failure.

Individuals

Project participants also include individuals, such as a prospective homeowner, who initiates the process of hiring designers and builders to develop a custom home on a piece of land. Also,

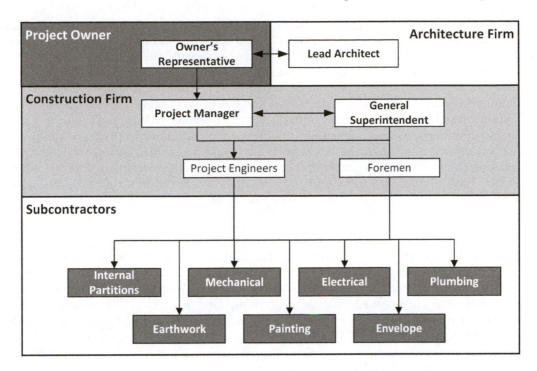

Figure 3.3 Project Organizational Chart

most projects are developed by employees of organizations for other organizations, which are represented by their own employees. Each project will have a different team of individuals that will operate within the contractual framework under a web of relationships that identify the *duties* associated with the job title, as well as the *responsibilities* and *authority* associated with the hierarchical position within the project.

For example, if a university wants to build a new training facility for its football team, one of the university employees in the capital project office will be assigned the duty, responsibility, and authority to serve as the owner's representative on this project. He or she will interact with the lead architect working for the architecture firm that has been hired to design the training facility. Later, the owner's representative will also interact with the project manager of the selected builder of the facility. Working closely with the general superintendent, the project manager has the ultimate duty, responsibility, and authority to make decisions for fulfilling the project objectives, as stated in his or her company's contract with the university; in addition, he or she will delegate duties, responsibilities, and authority to the project engineers as required. The building team will also oversee work performed by the subcontractors' employees. Figure 3.3 shows a simple project organizational chart.

Each of the businesses participating in the delivery of a built environment project relies on its employees. Here, we will outline some of the more common individual roles within each of the major participants.

Project owner's employees and agents

When a project owner frequently develops built environment projects, its staff may include individuals with expertise in one of the built environment disciplines. One of these employees will lead the owner's organization and act as the *owner's representative* with external entities, including coordination with AHJs, insurance companies and sureties, and design and construction firms. However, smaller owner organizations or individual owners usually rely on the external services of a design firm and/or an *agency construction manager* to serve the role of owner's representative. When both a design firm and an agency construction manager are employed, the design firm serves as the owner's representative during the initial phases of the project and the agency construction manager becomes involved during pre-construction and sees the project to its completion. The internal or external owner's representative often serves multiple roles, including (a) serving as the point of contact between the organization he or she works for and the other participants of the project, and (b) reporting to the upper management of the organization he or she works for. The owner's representative is sometimes called the *owner's project manager*. In public works, this individual may be called the *resident engineer*, as in the case of a state road project.

Design firm's employees and consultants

A design firm employs—or temporarily secures through consulting—individuals, including licensed designers in the specialty discipline the design firm operates in. For any given project, it is common for a design firm to act as the lead and hire other design firms as consultants in the design specialties that are needed. Each design firm involved in a project will usually have individuals serving three roles: a *designer of record*, a *lead designer*, and additional *design staff*, including design architects, engineers, and drafters. Although the lead designer often serves as the designer of record, two different individuals could serve in these roles. The designer of record will be the licensed individual who files a building permit request. The lead designer will be the individual who is charged with overseeing and managing a wider design staff toward project completion. This individual is sometimes called the design project manager.

Construction firm's employees

Construction firms employ many management specialists in addition to construction labor. Construction firms employ individuals with different levels of education, numerous job titles, and varying levels of authority. Though the title of *project manager* is well known inside and outside of the industry, most built environment projects are too complex to have only one individual in charge of the overall project success, field operations, relationships with client, communications with AHJs and other project participants, and project documentation handling. These duties are instead carried out by many individuals.

Construction firms frequently appoint a *project executive*, whose main duties include building long-term relationships with other project participants, improving the firm's reputation, and reporting to insurance and surety agents and to upper management or stockholders. This individual, also referred to as *officer-in-charge*, is usually someone with deep knowledge of the construction sector, including competitors, because he or she is expected to forecast how project actions may affect the future business of the firm beyond the completion of any given project. This individual will not customarily be located on one project site. Instead, he or she will visit all of the projects that are assigned and will also be called on to interact with project executives in the other project organizations when the project and/or field management teams on each side have not been able to come to an agreement on a given issue. For example, a commercial building general contractor will assign a project executive to a new project with the task of looking at the project success and also at enhancing the firm's reputation with the project owner to secure future business while maintaining successful long-term relationships with specialty contractors or design firms that the firm strategically partners with. Job titles for project executives serving in this project-oversight role vary across construction firms and include CEO, COO, account executive, senior project manager, and vice-president of construction operations.

Project management staff usually operate under the project executive. Since the project executive is rarely project-based, construction firms usually appoint a *project manager*, whose main duties and responsibilities are to achieve project success within the contractual requirements and, eventually, propose changes to the contractual requirements if they are needed to achieve a more successful outcome. This individual acts as the firm's lead individual on the project. When the firm's scope is too small to justify the full-time allocation of an employee, the project manager may be in charge of more than one project and may be located in the main office. This is the typical scenario for project managers working for specialty contractors that perform only a limited portion of a project's scope or for general contractors on smaller and tenant-improvement projects.

Field operations require a completely different skillset from project management, one that is unique to construction materials and methods. Construction firms usually staff their projects with a field manager who is in charge of field operations. The job title used by general contractors for this individual is usually *project superintendent* (or simply *superintendent*), or *general superintendent* to identify the lead field manager when more than one superintendent is employed on a project. For example, a heavy-civil general contractor may staff a large highway project with a general superintendent and additional superintendents for each of the road segments and day and night shifts. Similarly, a building general contractor may staff a large building project with a general superintendent and additional superintendents for major work items, such as structures, finishes, or envelope. When the firm's scope is small, a *general foreman* (or simply *foreman*) will serve this role. This is the typical scenario for specialty contractors who perform only a limited portion of a project's scope.

Whereas the project executive is involved in the project only periodically, the project manager and superintendent navigate the project's day-to-day activities. The relationship and authority of these two individuals change depending on the firm's culture, the individual's experience and knowledge, and the project type. Two scenarios are prevalent. Some projects are organized with a superintendent serving in a subsidiary position to the project manager. This scenario is feasible when a project manager with extensive field experience is assigned to the project. On other projects, the two individuals share or divide duties depending on their background, so they may operate at the same level, even if with a slightly different focus. This scenario is feasible when the pair's knowledge and experience complement each other, such as when a knowledgeable project manager may lack field experience relevant to the project, which is then instead provided by the superintendent.

The project manager often relies on one or more *project engineers*, who are delegated certain duties and report to the project manager or superintendent. For instance, a building project can have project engineers for overseeing structure, envelope, finishes, and building systems. These project engineers would oversee the delivery of these work items. Most project engineers do not carry job titles specific to their scope, but they may be layered by experience. Job titles frequently used include *project engineers* to identify entry-level assignments, *field engineers* to identify entry-level assignments that strictly relate to field activities, *senior project engineers* to handle advanced assignments, including the management of subcontracts, and *assistant project managers* to identify individuals receiving on-the-job training (OJT) to become project managers. A few specialty project engineering positions have appeared in the industry, including:

- *safety engineers*, project engineers who are fully vested in construction safety;
- *mechanical, electrical, and plumbing coordinators (MEP)*, project engineers on building projects who are fully vested in overseeing MEP specialty designers and contractors;

- *building information modeling (BIM) engineers*, project engineers who handle modeling and visualization of project documents; and
- *sustainability engineers*, project engineers who handle contractual requirements specific to sustainable projects.

Note: for the rest of this book, we will simply utilize "project engineers" when referring to all of these individuals.

Obviously, construction firms also employ construction labor for scope items to be self-performed. Labor is usually organized into crews. Each crew is led by a *foreman*, whose main duty and responsibility is to direct and lead craftsmen and report to a superintendent. Usually, a foreman also performs work alongside his or her crew. On large projects, a *general foreman* may be charged to oversee multiple crews. Craftsmen usually include a mix of *journey-level craftspeople* and *apprentices*. Apprentices are entry-level workers who are learning a trade through OJT under the supervision of other highly skilled workers. Journey-level craftspeople are experienced workers who have been recognized to be fully qualified to perform the skills of their trade through a process, the formality of which varies across jurisdictions and trades. They have learned their trade usually through completion of an apprenticeship or training program, including classroom time, practical work experience, or a combination of both. Journey-level workers can work without full-time supervision. Some trades also identify *master journey-level craftspeople* to recognize journey-level workers with higher levels of experience and skills.

Summary

The delivery of built environment projects requires the interaction of many businesses, organizations, and individuals. Major organizations directly participating in built environment projects include project owners, design firms, and construction firms. However, other organizations and individuals often perform important and crucial roles, including land surveyors, insurance companies, sureties, and AHJs. Private organizations operate under specific business forms depending on how ownership and control relate to each other. Proprietorships and partnerships give full control of the business to the owners who are, therefore, fully liable for the business enterprise. Corporations and LLCs separate ownership from control to limit the liability on the owners and make it easier to raise capital for the business. Organizations employ individuals with different duties, responsibility, and authority. Project owner organizations often rely on internal or external owner's representatives to oversee the project delivery. Design firms mostly employ licensed professionals who can apply for permits and serve as designers of record, with support from other licensed design professionals. To execute the project, construction firms employ many individuals with numerous job titles to fill their project organization needs, including project managers, superintendents, foremen, and project engineers.

Review questions

1. Who is a journey-level craftsman?
2. Who is an apprentice?
3. Who is an MEP coordinator?
4. What are the characteristics of a general partner?
5. What is the liability exposure of a proprietor?
6. Why would a project owner request a payment bond from the general contractor?
7. Among a general contractor's employees, who is mainly responsible for setting up the jobsite, including utilities and trailers?

Exercises

1. An owner requests a general contractor to secure a performance bond for its scope of work. Sketch a diagram identifying which is the PRINCIPAL, which is the SURETY, and which is the OBLIGEE under this situation.
2. Without looking ahead to the next chapters, list five responsibilities of a GC's project manager.
3. In addition to the items listed in question 7 above, provide five other responsibilities of a GC's superintendent.
4. List ten different and unique types of construction craftsmen who a project engineer may work with on a typical commercial building project.
5. Draw a project organizational chart for a building project that includes the following individuals: a project executive, a project manager, a general superintendent, a safety engineer, an MEP coordinator, a project engineer and a superintendent for structure, a project engineer and a superintendent for envelope, and a project engineer and a superintendent for finishes.

Reference

National Fire Protection Association (NFPA) (2017). NFPA 70: National Electrical Code, 2017 Edition. Retrieved September 26, 2017, from www.nfpa.org/codes-and-standards/all-codes-and-standards/list-of-codes-and-standards/detail?code=70

4   Project delivery

Introduction

As discussed in the previous chapter, projects have their own life cycle. All built environment projects follow a similar life cycle and involve most of the same types of participants, although phases may overlap and participants may interact differently depending on the owner's approach to deliver a project. The owner's strategy to finance a project is also relevant, because it may affect how a project is delivered.

The owner's initial plan on how to deliver the project is called the *project delivery method* (PDM). As briefly stated by Gibson and Walewski:

> [...] a project delivery method [...] defines the relationships, roles, and responsibilities of project team members and the sequence of activities required to complete a project
>
> (Gibson and Walewski 2001; p.1)

At the onset of any project, an owner has two options to provide services to the project, self-performing them or contracting them out. The higher the number of contracts used to deliver a project, the higher the need for coordination to keep everyone aligned to the same project objectives. Ideally, the simplest scenario for an owner would be to not have any contracts and to self-perform all of the work themselves, but this would require the owner to design and construct the project with their own staff. This scenario—sometimes called project delivery through force account—is extremely rare and will be described later in this chapter. Most built environment projects are instead delivered through several contracts for services, each following a similar contract delivery cycle to that shown in Figure 4.1a.

When a service is contracted out, the owner will need to clearly convey their expectations through a written contractual agreement that will describe the contract's scope of work and may package more than one service. However, this document is usually enforceable only if the other party had the opportunity to negotiate and approve its terms. Typical contractual agreements for architecture–engineering–construction (AEC) services will be discussed in the next chapter. Contracting out may occur in different ways, but it usually involves procuring external expertise for the project, which can be represented as a cycle that is repeated as many times as needed to deliver the whole project, as shown in Figure 4.1b. The selected PDM will dictate if each iteration will occur sequentially or concurrently to other iterations. Concurrency increases the need for coordination among contracted providers.

Consequently, the owner's decision to go with a specific PDM strongly affects the role and responsibilities of all project team members and how they will work together. This decision will also affect what portion of the life cycle will be affected, how many external entities will be involved in delivering this portion, and when they will be selected and perform their services under a contractual agreement. For instance, if an owner decides to have two separate firms delivering

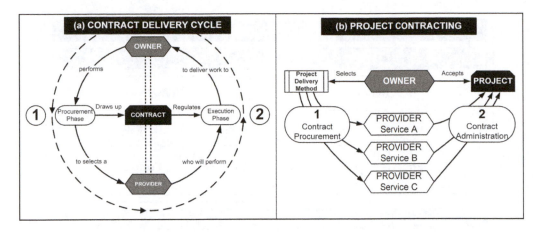

Figure 4.1 Project Contracting and Contract Delivery Cycle

design and construction services, each firm will enter into a separate contract with the owner to deliver their scope of work. If the owner decides to have the design completed before construction begins, the builder will not be involved in the design, so there may be more risk that design flaws will affect the execution of the construction activities.

Project delivery is becoming increasingly important as new ways to deliver and finance construction projects have been introduced, experimented with, and modified. This chapter introduces vocabulary used by industry practitioners to discuss project delivery and describes many common project delivery methods used in the industry. Moreover, it provides a structure for evaluating what project delivery method may be in place on a specific construction project and for understanding if and how a project delivery deviates from one of the more common PDMs.

This chapter intends to provide a brief introduction to this topic so that readers can gain a further understanding through examples in subsequent chapters and observation of actual industry projects. Moreover, this chapter does not discuss how project delivery affects subcontracts, which are used by prime contractors to procure the services and materials necessary to complete the scope of work. Subcontracts are described in Chapter 19.

Project delivery methods

Whereas some project delivery methods expand through construction into the operations and maintenance phases, the most common project delivery methods are (1) limited to the initial portion of the life span that ends at completion of the construction phase, and (2) constrained by how design and construction services may be packaged into contracts and the timing of these contracts. Traditionally, project delivery methods have been classified by their contractual structure, which is visually represented through a hierarchical diagram, also known as the project framework. This section describes several delivery methods and their project frameworks, as shown in Figure 4.2. However, new ways to deliver and finance construction projects are often introduced, which modify these methods and result in hybrid approaches.

Separated contracting of design and construction services

The first group of PDMs separates the delivery of design and construction services to the project, either by sequencing or overlapping them.

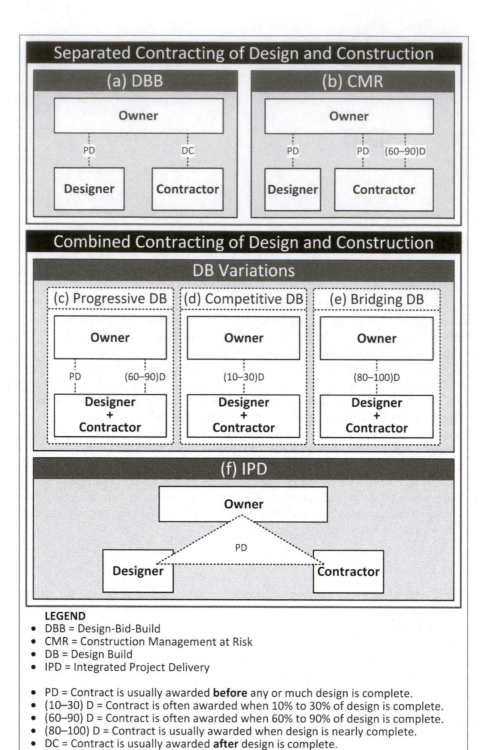

Figure 4.2 Timed Contractual Frameworks by PDM

Design–bid–build (DBB)

Under the DBB method, design and construction services are delivered to the project separately and, mostly, sequentially as shown in Figure 4.2a. First, the owner designs the project with in-house staff or enters in an early contract with a design firm that will produce a design package to be used as the basis for pricing construction services. Then, construction contractors are asked to review and price the project based on the design. In the public sector, this review and pricing phase is called the bid phase because laws for public work projects often require construction contracts to be awarded to the lowest bidder. Therefore, the design and construction contracts are usually timed in a sequential fashion. As a result, the construction company that will build the project is usually brought on board once most of the design decisions have been made.

This approach to delivering projects is often called the traditional method because it was the prevalent approach for delivering public projects in the 20th century and is still the most common approach to delivering heavy-civil public projects. In the private sector, owners are legally entitled to use the delivery method that they prefer. Still, DBB is often the method of choice because of the design community's familiarity with its processes and the legal community's reliance on its contractual agreements.

DBB example

A small company needs a new office building. This is a big investment for this firm, so its management prefers to evaluate all design options and finalize the design for the new office building before committing more funds by hiring a general contractor, who will procure materials and specialty trades to build the project.

- This owner does not have in-house design expertise, so they hire an architecture firm (the provider of design services) to help evaluate design options and produce a design package. This will include plans and specifications that can be used by the owner to understand that they are getting what they need and by a contractor to price the costs for constructing the building as designed. This firm brings its own engineering consultants for designing the foundation, the structure, and the building systems. [Note: Owners with in-house design expertise may not hire an external firm.]
- During the design phase, the owner, who does not have in-house construction expertise, selects and hires an agency construction management (ACM) firm (the provider of construction management services). The firm can help in selecting a general contractor once the design is completed and supervise the execution of the project. Also, the ACM will review the design to identify potential issues with its constructability and provide suggestions before the design is used as the basis of assumption for pricing the project. [Note: Owners with in-house construction management expertise may not hire an external firm.]
- Once the design is completed, the owner and the ACM (if any) work to identify a general contractor (the provider of construction services) who will price the project based on *what* needs to be built, *how* it needs to be done, and *when* it needs to be ready. The *what* is usually defined in the scope of work, which can be identified by reviewing the design drawings. The *how* defines methods and quality requirements, which can be identified by reviewing the project specifications. The *when* may be stated by the owner in their solicitation and will affect how the contractor will develop their schedule and select and assign resources to the project. These pieces of information will be discussed in more depth in Chapter 5 because they constitute the foundation of a construction contract.

Construction management at risk (CMR)

Under CMR, the owner contracts separately for design and construction services, which is similar to DBB, as shown in Figure 4.2b. However, the timing is different because the CMR is usually retained early on to allow early contractor involvement, which means the CMR will have an opportunity to review and comment on the design before it is finalized.

CMR evolved from the common use of agency construction managers in DBB. Whereas an ACM is simply a consultant to the owner, a CMR, as defined by the Construction Management Association of America (CMAA), is a firm that

> *[…] acts as [a] consultant to the owner in the predesign and design phases, but as the legal equivalent of a general contractor during the construction phase. When a CM is bound to a price, either fixed or a GMP, the most fundamental character of the relationship is changed.*

(CMAA 2017)

This change in role occurs when the CMR assumes a risk by committing to deliver the project according to the contractual requirements typical of a construction contract between an owner and a general contractor. Once the role changes, the owner–CMR relationship changes because the construction manager begins to protect themselves. For this reason, a CMR is also called a general contractor as a construction manager (GC/CM) or a construction manager as a general contractor (CM/GC). This is the two-step nature of CMR, with the second phase usually starting at the 60–90% design complete range, as shown in Figure 4.2b.

CMR example

A developer is planning a new high-rise tower to include underground parking, retail at ground level, and an upper-story hotel and apartments. Operators for the retail and hotel spaces have not yet been identified. This is the largest project to date for this developer, so their management prefers to retain some control over the design. Still, time is money, so the developer would like to overlap the design and construction activities without waiting for a hotel operator to decide details on that portion of the building. Therefore, the developer decides to use a CMR approach, where an architect and builder are selected early on and work together to further the design to a level of detail that can be priced.

- This owner does not have in-house design expertise, so they hire an architecture firm (the provider of design services) that will help understand what they are getting and work collaboratively with the builder to further the design to a level of definition that can be priced. They then continue to work collaboratively to complete the design and assist the builder during the construction.
- Soon after the designer is selected, the owner selects and hires a builder under an ACM contract (the provider of construction management services). This builder will work collaboratively with the designer to further the design to a level of definition that the builder can price.
- Once the architectural and structural design reaches around 60% completion, the owner and builder reach a price agreement on the structural scope and sign a contract that locks the price of the structure and establishes prices for the envelope, building systems, and finishes. Prices for the remaining scope items are tentative and frequently called allowances. The builder is now acting as a CMR and can move forward to finalize the structural design and begin the construction, which they will self-perform.

- Once the remaining design packages reach about 90% completion, the owner and builder reevaluate the allowances and adjust them upward or downward depending on the new available information. As soon as the owner and builder reach an agreement on each scope item, they lock the price. Once the price is locked, the builder can move forward by either self-performing or contracting out each of these scope items.

Combined contracting of design and construction services

Other PDMs combine the delivery of design and construction services for the project either by packaging them into a single contract or by establishing a multi-party contractual environment.

Design–build (DB)

Under DB, the owner enters into a single contract with an entity that will design and build the project. Thus, this contract will combine design and construction services. Acting as the single point of responsibility with the owner, the design-builder could be either a joint venture (JV) of firms or an individual firm that will subcontract out some of the services. For large projects, it is common that several construction firms will joint-venture by jointly constituting a new company that will be dissolved once the project is completed. For most projects, it is common for a general contractor to act as the design-builder while hiring a design firm to provide design services. Firms may be requested to list their consultants and/or subcontractors as part of the qualification submittal. Under any approach, the line of communication between the owner and the design consultants or subcontractors goes through the design-builder once the project price is set.

Among owners, DB has become popular due to several advantages over DBB, including having a single point of responsibility for design and construction, accelerated delivery, increased collaboration between designer and builder, and incentives for innovation. Still, the timing of design–build contracts may vary significantly. In the rest of this section, the three most common DB variations will be discussed. One variation of DB is the engineering, procurement, and construction (EPC) method that is mostly used in the industrial and utility sectors. EPC is usually based on a reimbursable contract that is established early on. Moreover, delivering a project through EPC usually provides a more turnkey experience for the owner.

PROGRESSIVE DESIGN-BUILD (PDB)

Under PDB, the contract is awarded early on. First, the owner selects an entity to act as the design-builder. Since little to no design is available, the selection is prevalently based on qualifications. The DB entity will work jointly with all parties, including its design consultants and the owner, to evolve the design to the point that it can be priced. At this point, a new contract can be signed between the owner and the design-builder, as shown in Figure 4.2c. To this end, PDB's two-step nature makes it similar to CMR. PDB is the most recent variation of DB and was introduced to overcome some of the issues with price-competitive DB, including claims. Claims will be introduced in Chapter 22.

COMPETITIVE DESIGN-BUILD (CDB)

Under CDB, the design–build contract is awarded once a portion of design is completed by the owner. A common level of design definition to award a CDB would be between 10% and 30% design completion, as shown in Figure 4.2d. First, the owner designs the project with in-house staff or enters into a first contract with a design firm that will produce a preliminary design package to be used as the basis for identifying the project's scope. Since some design is available, the owner will use this design information to convey what they want to competing firms. Moreover, the availability of some designs allows potential design-builders to price the project and, therefore, the selection also incorporates a price competition aspect. Sometimes, competing firms are requested to take this early design and move it forward enough to provide the owner with an understanding of where the project will be aiming if a certain firm is selected. This enhanced design is part of the firms' proposal together with information about costs, quality, safety, and schedule. Once a design-builder is signed on the project, the owner and the design-builder will work jointly within the contractual boundaries, so there is less owner control over the later design. Whether the design is self-performed by the design-builder or a service provided by a design consultant to the design-builder, the management of design services is substantially different from DBB. Whereas the project owner is highly involved in design activities under DBB and in the initial phase of PDB, involvement is limited under CDB to contractually allocated responsibilities for design quality control and assurance (QC/QA). Any further involvement results in potential change orders and affects the initial CDB contract price. Change orders are discussed in Chapter 20.

BRIDGING DESIGN–BUILD (BDB)

Under BDB, the design–build contract is awarded later than in the previous two processes. Whereas BDB is very similar to CDB, as shown in Figure 4.2e, this method relies on a set of advanced design documents (bridging documents) that constitutes the basis of the contract. Therefore, the owner's design consultant will enhance the level of design definition beyond what is usually achieved under CDB.

Integrated project delivery (IPD)

IPD is the most recent approach to project delivery and is based on lean construction principles. As defined by the American Institute of Architects (AIA), IPD is

> [...] *a project delivery approach that integrates people, systems, business structures and practices into a process that collaboratively harnesses the talents and insights of all participants to optimize project results, increase value to the owner, reduce waste, and maximize efficiency through all phases of design, fabrication, and construction.*

(AIA 2007; backcover)

From a contracting standpoint, IPD substantially differs from all other project delivery methods in the fact that it relies on one multi-party contractual agreement. All the contractual lines in the DBB, CMR, and DB forms are based on two-party agreements, whereas IPD contracts are multi-party. As a minimum, they include the three main parties—the owner, the designer, and the general contractor—but they can sometimes incorporate additional parties, such as specialty designers and contractors or vendors. A basic framework for IPD is shown in Figure 4.2f.

Project delivery dilemmas

Project delivery terminology

Language used to describe project delivery often lacks consistency. Therefore, the terms *project delivery*, *procurement*, and *contracting* are sometimes used interchangeably. To limit confusion on this topic, we will refer to *project delivery* as the act of achieving project objectives, to *procurement* as the act of purchasing external services and materials necessary to deliver a project, and to *contracting* as the act of establishing a contract for services and materials. Similarly, different ways to deliver projects can be interchangeably called project delivery methods, systems, routes, approaches, or strategies. Although the authors of this book consider the words "methods," "systems," "routes," and "strategies" as equivalent when used in reference to project delivery, "delivery method" will be used throughout the book.

Contracting versus self-performing

At the onset of any project, an owner has two options to provide services to the project, self-performing or contracting out. Under the first option, an owner would self-perform, which means the owner would carry out the work with their own forces. This approach is more frequent with design and construction management services because owners that deliver many projects often employ in-house design and construction management staff. When the same approach is used with construction services, the owner delivers the project via *force account*, which translates into the direct performance of construction work by use of labor, equipment, materials, and supplies furnished by the owner and used under their direct control. This approach is ideally the simplest because the owner would (1) not need to have any contracts, (2) retain complete control, (3) increase the speed of response, and (4) enhance the quality of work. However, this option would require the owner to (1) keep the workers employed, which is possible only if the owner continually delivers projects; (2) increase the payroll expenses, including base salaries, fringe benefits, and continuous training and education on evolving construction technologies; (3) own and maintain tools and equipment; and (4) fight the tendency for specialized skilled trades to lose the competitive edge as a result of operating in a non-competitive environment. These limitations and disadvantages of self-performance usually motivate owners to contract the services necessary for the delivery of a project, such as design and construction. Force account construction is sometimes used for repairs and maintenance. For instance, some cities own their own paving equipment and hire their own crews for repairing potholes and overlaying road pavement.

As previously mentioned, there are alternative project delivery methods from those discussed in this chapter. Similarly, some of the described project delivery methods can be customized to better fit an owner's needs and desires. One example of another project delivery method is *multi-prime contracting* (MPC). This method takes the DBB method of fragmentation of scope among contracts to a new level. Whereas DBB mostly splits design and construction scopes into two separate contracts, MPC splits, and eventually rearranges, the design and construction scopes into more than two contracts. An owner who selects MPC is deciding to self-perform most of the coordination that is traditionally contracted out to the design firm and/or the general contractor, as discussed in the following example.

MPC example

A school district in the urban area of a major U.S. city needs to create a new two-story above-ground primary school in its revitalized downtown to support the growing school-age children who reside in numerous new condominium buildings. This school district has significant experience in renovating its existing buildings through direct contracts with different trades, including mechanical, electrical, plumbing, and envelope. Many recent renovations were successful while being conducted on an accelerated schedule, so the district staff developed an ability to competently coordinate all of the trades that are needed to work in the same building at once.

- This owner does not have in-house design expertise, so they hire an architecture firm (the provider of design services) to help evaluate design options and produce a design package. This package will include plans and specifications that can be used by the owner to understand that they are getting what they need and by a contractor to price the costs for constructing the building as designed. This firm brings its own engineering consultants for designing the foundation, structure, and envelope. [Note: Owners with in-house expertise may not hire an external firm. Instead, they could use a multi-prime approach for design by hiring the engineering consultants individually.]
- Since the school district retains its own staff individuals with construction management expertise, it does not need to hire an external ACM firm.
- Once the design is completed, the owner's staff works to identify specialty contractors to deliver the individual components of the projects, including:
  - An earthmoving contractor who will prepare the land for construction,
  - A structural contractor who will build a shallow foundation and structure according to the provided design package,
  - An envelope contractor who will fabricate and install the selected building envelope,
  - A mechanical and plumbing contractor who will design and install heating and plumbing systems,
  - An electrical contractor who will design and install the electrical system, and
  - A finishes contractor who will frame and paint, lay floor covering, design and install cabinets, and install doors and furniture provided by the owner.
- During the construction execution, the owner's decision to not hire a general contractor translates into being directly responsible for self-performing some of the general contractor's responsibilities, including the coordination and supervision of the multiple contracts with their specialty contractors and lead designer.

Public versus private ownership

Another substantial difference in project delivery is the type of ownership. Private owners are usually legally entitled to use whatever delivery method they prefer. Therefore, the private sector utilizes a larger variety of delivery approaches, including hybrids of the traditional project delivery methods. On the other hand, public owners develop and deliver projects with taxpayer's money, so they are constrained by the laws, codes, and/or regulations in their jurisdiction. For instance, each state in the United States usually regulates public works through the state code, which also identifies which delivery methods are allowed for use on public works and which are not. Therefore, this code constrains public agencies within the state in their selection of project delivery method.

Project delivery features

As discussed earlier, increasing numbers of project delivery methods are used in the built environment, including hybrids of the ones discussed here. Whereas the public sector is constrained by laws, each private owner could hypothetically develop their own project delivery method or a multitude of them. This prompts a few questions:

- What are the important features of the delivery method for the project that a contractor will be working on?
- How are they recognized?
- How will they affect the work of a project engineer?

Project delivery methods can be differentiated by several features, including:

- the project life span, which identifies the period of facility life covered by the project delivery;
- the risk allocation method, which identifies the degree to which owner's transfer risks to industry providers;
- the contract packaging method, which identifies the degree to which contracts for different project services are combined; and
- the participation of industry providers, including construction firms, to the project financing.

The following examples are provided to help recognize delivery features in projects:

- When an owner decides to deliver a project through the design–bid–build method, they are also deciding that the project will end at construction close-out. Under DBB, the owner allocates many of the risks associated with construction to a general contractor. This method does not combine contracts for design and construction services. Finally, the designer and the contractor are not usually expected to contribute to the financing of the project beyond the invoicing cycle.
- Similarly to DBB, use of design–build for a project results in the project ending at construction close-out. Under DB, the owner allocates the risks associated with design and construction to an individual entity, the design-builder. Therefore, this method combines the design and construction services into a single contract. However, the design-builder is not usually expected to contribute to the financing of the project beyond the invoicing cycle.
- When an owner decides to deliver a project through the design–build–finance–operate (DBFO) method, they are also deciding to extend the project life beyond construction close-out to include a certain period of operation. The DBFO method allocates the risks associated with design, construction, and operation services, in addition to allocating a variable amount of the risks associated with financing. Finally, this method combines contracts for design, construction, and operation services. It also uses finance methods that include a funding component acquired by industry providers.

Project finance features

As previously stated, the owner's strategy to finance a project is relevant and may affect how a project is delivered. Therefore, a definition of project financing and a conceptual knowledge of the features of the approach for financing a project are relevant. A *project finance method* is defined as a system for acquiring or providing funds from different sources and combining them for financing a project during its delivery. Project finance methods can be differentiated by:

- the project life span, which identifies the part of the life of the corresponding facility that is financed;
- the types of financing sources that provide funding to the project (e.g., owner, lenders, or contractors); and
- the types of financing vehicles that are used (e.g., equity or loan).

The following examples are provided to help recognize the financial features in projects.

- A cell phone operator is granted a governmental license to build and operate a cell phone network in a new country. The operator lacks the $5 billion necessary to deliver the network. Therefore, they enter into a contract with a worldwide telecommunication company that will plan, design, build, finance, and maintain the network. Acting on behalf of the project owner—i.e., the cell phone operator—the telecommunication company will oversee all project phases and will hire and pay design and construction firms while managing these contracts. Once the network is operational, the telecommunication company will be reimbursed for its financing through a percentage of the revenues from the network subscribers over the initial five years of operations. At the end of this period, the cell phone operator will pay any residual amount, including interests, in a balloon payment.
- A public mass transit agency needs to fulfill a transportation need for the residents in its region. Although planning has shown the agency how it may fulfill this need, funding for its design and construction is lacking. Thus, the agency puts a proposition for a temporary tax increase in local taxes on the ballot.
- A state road transportation agency needs to envision how to deliver a new statewide corridor that will facilitate freight transportation while reducing traffic congestion for motorists. Several technical options are available that include only additional roads or a mix of roads and rails, but the agency lacks funding for their delivery. Thus, the agency initiates a public competition among private entities to suggest both a technical and financing option for achieving the agency's objectives. Everything is on the table, and private parties can propose any of the technical options, as well as a financing scenario.

Project engineering applications

Role and responsibilities

The role and responsibilities of construction project engineers will be significantly different depending on the PDM because the scope of work and the time of involvement in the project life cycle will be different. The role and responsibility of a project engineer on a DBB project will mostly rely on a fully designed scope of work. Therefore, the amount of pre-construction activities will be limited to reviewing construction documents for inconsistencies and coordinating with subcontractors to facilitate the production of shop drawings. On the other hand, project engineers working on DB and CMR projects may need to be fully involved in a broad range of pre-construction services to forecast issues and proactively address them. Any shortfall in this reading-the-crystal-ball exercise will result in problems that the project engineer will need to solve later.

Communications

PDMs that rely on overlapping phases of the project life cycle require more intense communication among all field and office project parties. The general contractor's project engineers serve an important role in ensuring that this communication occurs daily and regularly. Therefore, project engineers working on projects where communications are expected to be more intense should be well-versed in understanding how to better communicate with others while adapting their own communication style to that of others. Communications may have contractual implications, so project engineers should be familiar with the contract(s) they are operating under and the subcontract(s) they are overseeing. Since contracts are often contextual to the project delivery method, this contract review will require tuning of a project engineer's knowledge base.

Summary

In construction, the project delivery method is the owner's plan for how to deliver their project. A good understanding of the adopted PDM provides a navigation compass to project engineers and project managers alike. Today, certain industry sectors and their owners are quickly evolving and pushing the boundaries of project delivery. Whereas, in the 1950s, a large majority of new project engineers could think they would work for most, if not all, of their career on DBB projects, project engineers today are expected to be adaptable to the owner's approach for delivery. This chapter was just a brief introduction to the topic of project delivery and has hopefully sparked interest in the reader. The forthcoming chapter will continue to build upon these concepts by discussing procurement and contracting.

Review questions

1. What is a project delivery method?
2. What project delivery method is frequently called traditional?
3. What project delivery method is most commonly used to deliver heavy-civil public projects?
4. How might the role of a project engineer be different on a CMR versus a DBB project?
5. Which of the delivery methods described in this chapter are most applicable to each of our case studies included in Appendix A?

Exercises

1. Sketch the contractual framework for each of the following project delivery methods:
   a. Design–bid–build
   b. DBB with agency construction management
   c. Design–build
   d. Design–build with agency construction management
   e. Design–build–maintain
   f. Construction management at risk
2. Review the following case examples, sketch their contractual frameworks, and suggest the project delivery methods they may refer to:
   a. An owner enters into a contract with a joint venture that includes three companies (A, B, and C) and provides them with a program for a roadway that includes schematic design and incomplete specifications. Under the joint-venture agreement, Company A will be responsible for design, Company B will self-perform most construction activities and oversee subcontracts, and Company C will maintain the roadway for 15 years after completion.
   b. An owner hires a firm to design a facility and then hires another firm to provide construction expertise during the design process. Once 60% of design has been completed, the second firm signs a guaranteed maximum price contract and takes on most of the construction risks.
   c. An owner designs a facility with in-house personnel and then selects the company that provides the lowest price for the construction based on the provided design package.

    d.   An owner enters into a contract with a firm that will completely design, coordinate, and manage the construction process for the facility.

    e.   An owner fully designs an office building with in-house personnel and then selects a builder who will perform earth and structural works; three specialty contractors who will design and build mechanical, electrical, and plumbing systems; and another builder who will coordinate finishes through a series of subcontracts.

References

Gibson, G. E., & Walewski, J. (2001). Project Delivery Methods and Contracting Approaches Available for Implementation by TxDOT (Project Summary Report No. CTR 2129-S). Austin, TX, USA: University of Texas at Austin.

CMAA (2017). CMAA Glossary. Retrieved September 26, 2017, from http://cmaanet.org/glossary

AIA (2007). Integrated Project Delivery: A Guide. Retrieved September 26, 2017, from https://info.aia.org/SiteObjects/files/IPD_Guide_2007.pdf

5 Procurement and contracting

Introduction

Whereas the delivery method selected by a project owner identifies the overall framework for delivering its project, *procurement* and *contracting* are two additional and necessary aspects of project delivery that will affect the project execution. To deliver a built environment (BE) project, several project participants will be selected through *procurement* to operate under contracts that will be formed and negotiated through *contracting*. Contracts can be categorized into *prime contracts*, *subcontracts*, and *other contracts*. Figure 5.1 provides examples of each for a hypothetical design–bid–build project where the project owner is assisted by an agency construction manager.

Prime contracts are agreements between the project owner and another party. In the figure, the three prime contracts are between the project owner and the design firm, agency construction manager, and general contractor (GC).

Subcontracts include all contracts following the hierarchical structure of the project delivery method that do not involve the project owner. In Figure 5.1, contracts between the GC and each of the electrical contractor, the envelope contractor, and the joint-venture entity between the mechanical and plumbing contractors are subcontracts for construction services. One of the main characteristics of subcontracts is that they mirror some of the language of the overarching prime contract to enforce coordination among project participants and promote their alignment to the overall project objectives. For instance, the payment section of the subcontract between the GC and the electrical contractor will probably follow the *pay when paid* approach. This mirrors the prime contract between the project owner and the GC, in that the electrical contractor should not expect payment by the GC before the general contractor receives payment from the project owner.

Most of the contracts on a BE project are either prime contracts or subcontracts. However, there are exceptions, which we will categorize as *other contracts*. As an example of other contracts, Figure 5.1 shows a joint-venture agreement between the mechanical and plumbing contractors that team up to sign a subcontract with the GC. Therefore, the plumbing and mechanical joint venture is brought to the project with a subcontract, but they hold a joint-venture agreement with each other for the duration of the project.

Procurement overview

Procurement is the act of purchasing the external services and materials necessary to deliver a project. Any time that a project participant needs materials or services, they will initiate procurement to select another project participant who will provide them. For instance, a project owner can initiate procurement to select a design firm that is qualified for this type of facility or infrastructure and, later, enter into a prime contract with the selected firm. Similarly, a GC can

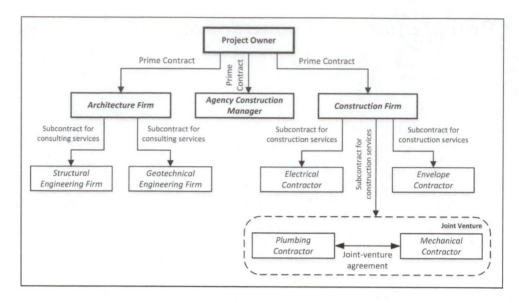

Figure 5.1 Types of Contracts

initiate procurement to select specialty contractors to help them achieve the project objectives by delivering portions of the facility within their specialty, such as electrical systems or mechanical systems.

Although project owners initiate procurement on a project by selecting design and construction firms with which to negotiate a prime contract, procurement is not a privilege of project owners. In fact, design and construction firms working under prime or subcontracts also rely upon procurement to purchase external services and materials through subcontracts. However, project owners can directly or indirectly influence the procurement of subcontractors.

Procurement selection methods

Several features shape the procurement of a contract. First, the decision of how to select a prospective project participant will influence the selection and is best accomplished upfront, which translates into identification of what *decision criteria* will guide the selection. Although any criteria could be used, criteria revolving around the dimensions of project success that were discussed in Chapter 2 are a good place to start. Therefore, cost, time, quality, and safety criteria are prevalent in any procurement decision. However, the importance of each may vary across projects. For instance, a bank may base its selection of a design firm and a construction firm around quality and safety criteria when dealing with the construction of its new headquarters. However, the same bank may base its selection around price and time criteria when dealing with the construction of a new suburban branch office.

Another important feature of procurement is the *selection process*, which can be finalized in a single step or over multiple steps. This feature is often dependent on the owner's desire to manage competitiveness of the business environment for design or construction services. For instance, the bank will have more chance to purchase services at a lower price for its new suburban branch office when more construction firms compete for the contract, but this will also mean that the selection process may become overwhelming for the bank. Therefore, the bank could decide to perform a pre-qualification on the basis of specific criteria to shortlist the most credible firms before asking for more information from each of the finalists.

Moreover, when procuring a prime contract, some project owners may "deepen" their selection by asking competing general contractors to propose some of their subcontractors and then evaluating each GC together with its team. Project owners may elect to use this approach if some components of the project are particularly important to them and they want to have a strong say in the selection of specialty firms that will deliver that component. A similar outcome could be obtained by using multi-prime contracting delivery, in which each specialty firm operates under a prime contract. However, this delivery method increases the coordination burden on the project owner.

The remainder of this section describes several procurement methods with a focus on their selection decision criteria and selection process. However, new ways to procure construction contracts are often introduced and these methods become modified, which results in hybrid approaches.

Competitive bidding selection

Sometimes called *low bidding*, this procurement approach uses price as the main criterion to select construction firms. Projects delivered through design–bid–build delivery often rely on this procurement method to select a construction firm to operate under a prime contract. Competitive bidding is prevalent in public projects because it is expected to provide the lowest price through competition among firms, and it may even be required by law. Prime contractors may also rely on competitive bidding to select specialty contractors that will operate under subcontracts. The use of price-based competitive bidding to select design firms is discouraged by professional societies and rarely used to award major design contracts.

In a typical design–bid–build scenario, a design package is developed early on by the project owner or by their design consultants. The project owner will first accept the design package and then provide it to competing construction firms together with a proposed contract. Construction firms that decide to compete for the contract will use the design package as the basis for estimating the project cost. Then, they will review the proposed contract to evaluate its risk allocation, add their markup based on their risk considerations, and submit a bid price to the project owner. The process used to develop a bid price will be discussed in Chapter 17. By submitting a bid, a construction firm is usually committing to signing the contract at the bid price if its bid is the lowest. However, if the design is flawed or needs to be changed later, most contracts entitle the construction firm to a contract change by negotiating a *change order* with the project owner, as discussed in Chapter 20.

Competitive bidding is popular due to its simplicity. However, there are several risks associated with using only the price for selecting a new project participant. Variations to a simple selection of the low bidder exist to overcome some of these issues, a few of which are discussed below.

By adopting the competitive bidding procurement approach, the owner selects a construction firm based on a price, which is based on the design package. Therefore, the *quality of the design package* is often what makes or breaks the effectiveness of competitive bidding. Design flaws enhance the risk of conflicts during the contract administration and price escalations through contract changes. Several factors contribute to the quality of design and its *constructability* or *buildability*. First, the accuracy and completeness of information on existing site conditions provide a foundation for the design, so a design based on incorrect site information is often heavily flawed. The gathering of accurate site information is often difficult because extensive site investigations may be very expensive or too difficult to perform without disturbing the existing site conditions. Owners often address this issue by incorporating in the contract specific processes for handling risks associated with *differing site conditions*, which will be discussed in Chapter 20. Second, the experience and expertise of the design firm with the project type and location is crucial to forecast issues and differing site conditions. Therefore, owners rarely use competitive bidding to select design firms because selection of a design firm should not be based on price alone. Third, design is a human activity and the time available for completing the design package affects the risk for human errors. Since project owners bear most of the risks associated with defective design, they should refrain from accelerating the design schedule. Fourth, the level of definition in the design package affects how competing construction firms price project elements. An incomplete level of design definition forces competitors to make assumptions while developing their bid price. These assumptions may be the basis for later disagreement with the project owner. Therefore, competitive bidding should only be used when an accurate and complete design package can be provided to contractors, so that the number and magnitude of these assumptions is reduced. However, it is almost impossible to produce a 100% complete set of plans and specifications. Last, the use of construction knowledge and experience is important during design to identify issues that could impede the constructability of the facility or infrastructure as designed. A design package with significant constructability issues is expected to result in significant price adjustments because change orders are not priced competitively but are instead negotiated with the selected construction firm.

As clearly stated in the previous examples, selection of a construction firm on price alone ignores the other tangible and intangible factors that affect project success, and it may result in the selection of an unreliable firm. To address this issue, other preliminary steps or requirements are often added to the selection process. By extending the selection to another quantitative factor, time, some owners may adopt a price and time selection (also known as A+B bidding in the heavy-civil sector), in which competitors need to submit a project duration along with their bid price. This approach is quite useful for owners who would benefit from an early completion. Under this procurement process, each day of project duration is assigned a monetary value, and the bid price would be the sum of the estimated price plus the time monetary value.

To take into consideration other selection factors, some owners may pre-qualify firms before inviting them to submit a bid. The selection would still be based on price, but only firms meeting pre-qualifying requirements would be allowed to submit a bid. The pre-qualification step can be carried out on a project-by-project basis or periodically. Some owners maintain a list of pre-qualified firms, which is updated periodically. To stay on the list, firms will be evaluated based on their performance with the given owner, but they may also be required to provide information and references for projects carried out for other project owners. However, this approach may not make sense for project owners that do not frequently deliver BE projects. Under these situations, an owner can add a pre-qualification step by issuing a request for qualifications (RFQ) to interested firms. Then, the project owner will evaluate the qualifications and screen firms that will be invited to submit a bid. The pre-qualification step is used to evaluate firms on criteria other than price and to pre-qualify those firms that meet minimum requirements on criteria other than price.

Another approach to pre-qualify contractors is to ask them to bond the project. In this case, the project owner is outsourcing investigation on the reliability of the competing construction firms to sureties. These sureties will take some of the risks associated with a potential failure of the firms to perform (issuance of performance bond) or pay their subcontractors and vendors (issuance of payment bond). The role of sureties on BE projects was introduced in Chapter 3 and will be further discussed in Chapter 21.

Last, selection of a construction firm based only on price does not promote innovation or allow for facility-life considerations during construction. This issue can be addressed by incorporating contractual incentives for construction firms to propose changes to the design package that would add value or reduce costs. This is usually undertaken by including the opportunity to submit *value engineering (VE) change proposals* during contract execution. If accepted by the project owner, cost-reductive VE change proposals may result in some of the savings being shared with the contractor.

Similarly, project owners can provide for price adjustments to address value–adding VE change proposals throughout the construction process.

Example

A city's department of transportation is planning to solicit bids for a road project in an urban area. The three main project objectives are to get the best price, to complete the project within 150 calendar days, and to incentivize contractors who will reduce the impact of construction on road users. To this end, the owner adopts an A+B bidding procedure to select its prime contractor. Construction firms competing for this contract are invited to submit a bid amount and a project duration in calendar days. Bids that would envision a project duration longer than 150 days would be deemed not responsive and excluded. A daily monetary value of $15,000 will be assigned to each construction day based on an estimate of road user costs. Table 5.1 shows the submitted bids by four contractors.

Table 5.1 A+B Bid Summary

Contractor	Bid Amount (A)	Project Duration in Calendar Days (B)	Road User Cost	A+B Bid
A	$ 4,500,000	150 days	$ 15,000/day	$6,750,000
B	$ 4,870,000	110 days	$ 15,000/day	$6,520,000
	Contract Price			**Low Bid**
C	$ 4,700,000	125 days	$ 15,000/day	$6,575,000
D	$ 4,200,000	180 days	$ 15,000/day	Excluded

In this example, Contractor B has the lowest A+B bid, so a contract of $4,870,000 would be awarded to this contractor under the agreement that the project will be completed within 110 calendar days. To incentivize on-time completion, this contract includes a *liquidated damage* (LD) clause, in that final payment to the contractor would be reduced by an amount of $15,000 for each day of late completion.

Best value selection

This procurement approach is designed to align prospective firms with the project owner's objectives by asking them to develop a proposal that allows for best value selection for the project. This approach attempts to overcome some of the issues present in competitive bidding. Best value selection is more flexible because it can be adapted for use under any project delivery method. Also, by de-emphasizing the importance of price alone, best value selection adopts a multi-criteria selection. Other selection criteria can vary and include both qualitative and quantitative criteria. Examples of qualitative criteria include references, experience with the project type, and experience and qualifications of the proposed project staff, including the project manager and superintendent. Examples of quantitative criteria include price or fee percentage, target project duration, and safety performance indices.

Best value selection usually follows either a one-step or two-step selection process. Under the one-step process, competing firms are invited to respond to a request for proposal (RFP) that details the expected content of proposals and describes the selection process. Since all interested proposers will submit information at the same time, review of the proposals could be cumbersome. Therefore, this process is preferably used when the project is smaller or simpler, the information to be submitted is limited in scope, or less competitors are expected.

In all other situations, a two-step process is preferred. Under two-step selection, competing firms are first invited to respond to an RFQ by submitting information on their firms that would help the project owner identify which firms would have the best chance to achieve all of the project objectives at the best value. Then, the project owner will usually invite three to five shortlisted firms to respond to an RFP, when each firm will develop their proposal for delivering the project. As stated earlier, best value selection can be adopted on different project delivery methods. For instance, it could be used to replace a competitive bid for selecting a prime contractor based on a completed design. This approach would be similar to a competitive bidding situation with pre-qualification. The main difference would lie in the shortlisting of a given number of firms instead of the pre-qualifying of all firms that meet certain requirements. Still, best value is more frequently used to select project participants under design–build and construction management at risk delivery methods, as discussed in Chapter 4. Since best value still weighs price, competing firms are expected to provide a price as part of their proposal, which is highly dependent on the quality of the design. Therefore, best value selection is used when a fair amount of design has already been completed and can be used to provide a basis for estimating the project cost.

Qualification-based selection

Similarly to best value selection, qualification-based selection is highly flexible and can be adapted for use under any project delivery method. However, this selection process de-emphasizes the importance of price by adopting a multi-criteria selection in which the price criterion is minimally weighted, if considered at all. This characteristic results in the selection of project participants without a price being available for their scope of work. The resulting contract would therefore be based on other issues, such as cost-reimbursable terms.

Sole-source selection

This procurement approach is radically different from those previously described, because it excludes competition. Instead, a participant is selected in advance because they are the only one capable of providing the service or are preferred to satisfy the project needs for justifiable reasons, so competition is not needed. This is a common approach for selecting specialized project components. For instance, an equipment manufacturer can be sole-sourced to provide a new turbine for an existing electrical plant because it is the only manufacturer that produces a turbine of the right power and dimensions to fit in the existing facility. Similarly, the selection of a specialty contractor to install artificial turf for a high school football field may result in sole-sourcing the only firm able to provide the service for the selected material. The decision to use this non-competitive approach often requires justification, including investigations of potential competitors or prior good references. Unlike the previously described procurement approaches, the burden of this data gathering would be on the selecting party.

Contracting overview

Once a new project participant is selected, *contracting* is the process of establishing a contractual relationship for services and materials through the development of a written agreement expressing the expectations, responsibilities, and protections of each party. Contracting for construction services is substantially different from contracting for design services.

Traditionally, the process of establishing a contractual relationship for design services is often based on reciprocal trust between parties. In this case, the purchasing party trusts the technical expertise of the designer, whose selection is usually based on qualifications. Moreover, design contractual documents often create an agency relationship by authorizing the designer to act as an agent on behalf of the client for various decisions, such as permitting or design interpretation. Agency relationships are implicitly trust-based. Contracting for construction management and pre-construction services retains many of the elements of design contracting, including the agency relationship.

On the other hand, the process of establishing a contractual relationship for construction services is based on an *arm's-length transaction*, which legally means the two parties are working in their self-interest during the contracting and are free of reciprocal pressure or duress. As a result, construction contractual documents are expected to be thorough, grounded in the design developed at the date of the contract, and crafted to allocate risks among contractual parties. Design–build contracting retains many of the elements of construction contracting, but it often results in the production of contractual documents that more flexibly adapt to the project objectives. In the remainder of this book, only contracting and contractual documents for construction services will be discussed because they are the most relevant to the duties of a construction project engineer.

Construction contract documents

A contractual relationship for construction services and materials relies on a set of written documents. The main document is a *contractual agreement*. Through a legal concept called *incorporation by reference*, the construction contractual agreement is also used to incorporate the other contractual documents, such as:

- The *general conditions* document, which describes the operating procedures that the owner usually uses on all projects of similar type;
- The *special conditions* document, which describes the operating procedures that are unique requirements for the project;
- The *construction drawings*, which provide a graphical description of the project, including its geometrical information and often its materials; and
- The *construction specifications*, which provide a textual description of the project, including any additional information on its materials, quality acceptance processes, or other performance expectations. Construction specifications are usually organized according to some standardized classification system. For instance, the Construction Specification Institute (CSI) MasterFormat is a commonly used classification system for organizing data about construction requirements, products, and activities.

To incorporate other documents, the *contractual agreement* must refer to them, provide them as attachments, and clearly state that they are included as part of the contract, as shown in Figure 5.2. This approach allows for the organization of the rights and obligations of each signatory party into a set of documents that describes *what* needs to be built, *how* it needs to be done, and *when* it needs to be ready. The *what* is usually defined in the scope of work, which is summarized in the contractual agreement and detailed in the construction drawings, the *how* is the methods and quality requirements, which can be identified by reviewing the project specifications and conditions,

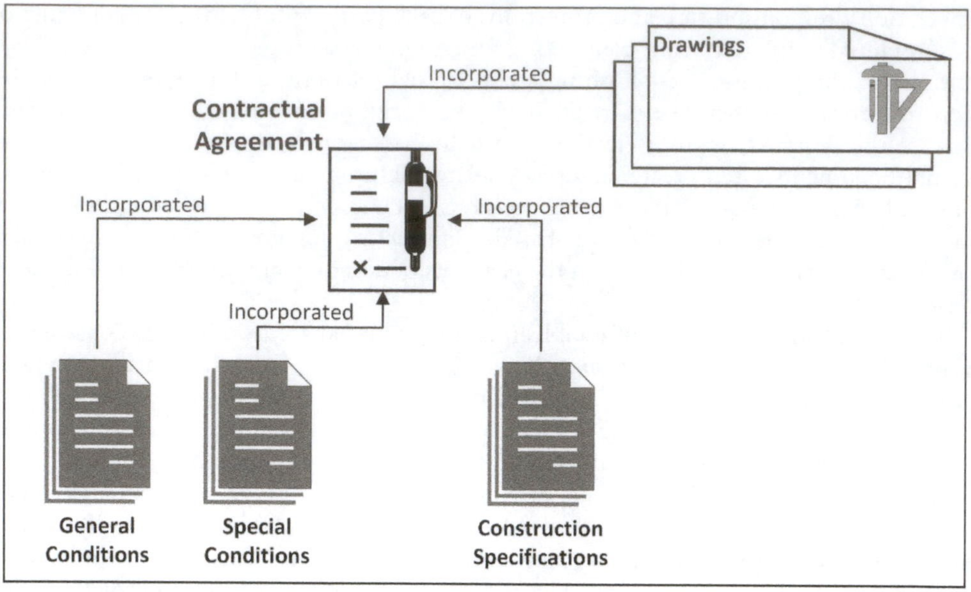

Figure 5.2 Construction Documents

and the *when* will be stated in the contractual agreement. As a minimum, a construction contract includes the information listed below:

- *Who*: The signatory contractual parties and any relationship to other parties;
- *What*: The work to be performed;
- *Where*: The location of the project;
- *When*: The time allowed to perform the project, including any financial disincentive for late completion (under liquidated damages);
- *How much*: The pricing method, including contract price, if known, or the process to evaluate costs to be reimbursed and fees to be awarded;
- *How to get paid*: The payment method, including processes to obtain progress and final payments; and
- *Other hows*: Processes to address issues that may arise during contract administration, including changes to the contract, disputes, unforeseen circumstances, design flaws, and termination of the contract.

Prime contracts

A contractual relationship for construction services and materials between a project owner and a construction firm relies on a set of documents that is usually drafted by the project owner and their consultants. The main document is a contractual agreement, also known as the *construction (prime) contract*. This document defines the rights and obligations of each signatory party, including the construction firm. Drafted by the owners, prime contracts for construction services are submitted to the selected construction firm for review and acceptance. During the review, the parties may decide to re-negotiate some of the contractual terms before signing the contract. Various factors, including the nature of ownership (private versus public) and the project delivery and procurement methods, affect whether there is room for negotiation and what can be negotiated. Private owners tend to provide more opportunities for negotiation than public owners do, because public owners can rarely negotiate anything that may have affected the selection. For instance, if a construction firm was selected by a public agency based on their price and time offer, the contract sum and the delivery time cannot be changed. Most public procurement laws would forbid negotiating terms that were used to make the selection or affect it.

Subcontracts

A contractual relationship between a construction firm and an entity other than the project owner also relies on a set of documents. The main document is a contractual agreement, also known as the *construction subcontract*. Similarly to a prime contract, this document should define exactly and explicitly the rights and obligations of each signatory party. Subcontracts for construction services are usually developed by the buyer of the services, who is frequently a prime contractor, and replicate aspects of the prime contractual documents to share some of the prime contract risks with the subcontractor. Subcontracting is discussed in Chapter 19.

Standard forms of agreement

Whereas large private organizations and public agencies often develop their own set of contractual documents, sets of standard forms of agreement are also available through various industry organizations. These sets include documents, sometimes improperly named *standard contracts*, which can be adapted to a specific project by filling in the blanks in the template. Some sets of standard forms of agreement are most frequently used by owners who do not use a proprietary set of documents:

- *AIA Contracts*, developed by the American Institute of Architects (AIA): Figure 5.3 shows the relationship between some of the AIA contract documents;
- *ConsensusDocs*, developed by 40 leading associations in the construction industry;

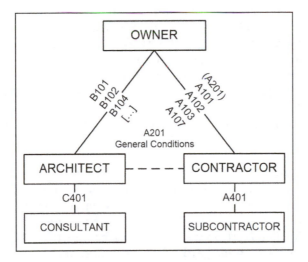

Figure 5.3 AIA Documents

- *NAHB Contracts*, developed by the National Association of Home Builders (NAHB) for residential projects; and
- *Engineers Joint Contract Documents Committee (EJCDC) Contract Documents*, developed by a coalition of three major organizations of professional engineers, namely, The American Council of Engineering Companies, The National Society of Professional Engineers, and The American Society of Civil Engineers – Construction Institute.

Types of construction contracts

Information to be contained in a contractual agreement is highly dependent upon the project. Therefore, contractual terms can vary greatly among projects, with the exception of the pricing and payment methods. In fact, pricing methods for construction contracts mostly fall within three types: (a) *cost plus a fee* (with or without a guaranteed maximum price), (b) *unit prices*, and (c) *lump sum*. The pricing method often affects the payment method. As with the project delivery and procurement methods, the owner also selects the project pricing and payment methods for prime contracts.

Cost plus a fee

This pricing method, also referred to as simply *cost plus*, prices the contract by adding up two elements. The first element is the *cost* to perform the contract work, which is the sum of the direct costs (the costs of construction, including labor, materials, subcontracts, and equipment that can be directly related to specific work items) and the portion of indirect costs that are not directly related to any specific scope of work but apply to the entire contract. The second element is the *fee*, which can be expressed as a percentage of the cost element, as a fixed amount, or through an incentive fee. Under a cost plus with incentive fee, the contract "provides for the initially negotiated fee to be adjusted later by a formula based on the relationship of total allowable costs to total target costs" (GSA 2017; subpart 16.405-1). The fee element allows the construction firm to charge the contract for the expected profit, as well as the portion of indirect costs that cannot specifically apply to the contract work but that represents the cost of doing business, also known as the *home-office indirect costs*.

The cost plus method offers advantages and disadvantages to both parties. Since the pricing is cost-dependent, the payment method will require the construction firm to show its costs through *open-book accounting*, which provides transparent access to cost information for the purchasing party but adds more bureaucracy to the contract administration. From their nature, cost plus contracts are especially advantageous when the design is not complete and for complex projects because they do not allow the contractor to hide contingencies. For instance, a project owner using this method to contract with a prime contractor would have information on all subcontracted work and prime contractor labor and material costs, but they would also need to review all invoices before processing a monthly payment. However, it also means that the contract price is unknown at the start, which creates uncertainty about the contract price so it cannot be used as a selection factor.

To address this price uncertainty, cost plus contracts can add a *guaranteed maximum price (GMP)* condition. When a GMP is added to a cost plus contract, the contractor is taking the risk of cost overruns beyond the GMP. Under a cost plus with GMP contract, the owner is agreeing to pay for the work up to the GMP value, which acts as a price cap for the owner. On the other hand, it is difficult to secure a firm GMP too early in the design phase because the contract work is still vaguely defined. If a cost plus contract is signed early on, for instance, under construction management at risk project delivery, the GMP cap would only be added later, once the design package has evolved enough to allow the contractor to price it. As a rule of thumb, the design package should reach 60–90% of design completeness before an accurate GMP can be developed. These percentages should not be considered in absolute terms for each item of the scope of work, as discussed in the next example.

Example

A developer secures a cost plus contract with a general contractor for a high-rise condominium project early in the design. Once the design package evolves to roughly 30% design completeness, the parties agree on a GMP for the sitework, earthwork, and structural scope, because these elements of design have already reached 60–90% completion. However, the envelope, building systems, and finishes are still vaguely defined. To budget funds toward these scope items, the parties estimate their price at the current level of definition. These tentative values are called *allowances* and are not intended to cap the price. Instead, they are used to guide design and to identify cost overruns early on.

Unit prices

This method provides a ballpark contract price by summing estimated prices for each scope item. These prices are obtained by multiplying the *quantities* associated with each scope item—estimated by the project owner—by the *unit prices*, which are provided by the contractor. The use of the unit price method is common on public works and heavy-civil projects, such as roads and utilities. Under this contract, the construction firm will provide a list of unit prices for various work items during procurement, such as cubic yards of excavation or linear feet of 6" drainage pipe. Since these prices are dependent on quantities, the project owner provides estimated quantities and bears the risk associated with quantity estimating errors, whereas the contractors bear the risk of pricing errors. This method is often associated with contractual adjustments for quantity variations outside a contractually set range. It is also often associated with a matching payment method where the project owner's inspectors evaluate quantities in place and compare them against billed quantities before authorizing progress and final payments. Since somewhat reliable quantities are necessary, design activities should be at an advanced stage at the contracting time, similarly to the lump sum method. However, the unit price method does not transfer the risk of erroneous quantity take-off to the contractor, which is beneficial for heavy-civil projects where quantities may be only tentative.

Example

A city wants to secure maintenance contracts for two of its neighborhoods to fill potholes in gravel roads with a compacted sand/gravel mix. The city selects a construction firm whose unit prices suggest the lowest total project price for the two neighborhoods. Each contract allows for renegotiation of unit prices when estimated quantities are off by ± 20%. During construction, these quantities are adjusted to the actual values, which affects the total price, as shown in Table 5.2.

Table 5.2 Unit Price Contract Summary

	Procurement			Actual		
	Quantities (Cubic Yards)	Unit Bid Price	Contract Price	Quantities (Cubic Yards)	Unit Bid Price	Contract Price
Contract 1	1,200	$ 600/cy	$ 720,000	1,100	$ 600/cy (bid)	$660,000
Contract 2	1,000	$ 600/cy	$ 600,000	1,400	$ 640/cy (renegotiated)	$896,000

Lump sum

This pricing method, also referred to as *stipulated sum* or *fixed sum*, prices the contract upfront. Therefore, this method is preferable for projects where the design is sufficiently complete, so that the construction firm can reasonably estimate project costs, but it has also been used to price design–build contracts based on incomplete design packages. However, the use of this method when the design is incomplete tends to inflate the project price by incorporating contractor's contingencies for risks that may not materialize. Whereas the lump sum price can cumulatively price all work to be performed, a *schedule of values (SOV)* is usually adopted to break the price down into work items and is used to identify progress payments through an assessment of the percentage of completion of each item. SOVs can be used under any contract type. The CSI MasterFormat is a commonly used approach to present a schedule of values.

Project engineering applications

The role and responsibilities of construction project engineers in procurement or contracting vary significantly depending on the project delivery method, because the scope of work and the time of involvement in the project life cycle is different. The role and responsibilities of a project engineer in procurement can include supporting the team in quantity take-off and pricing during bidding, contacting potential subcontractors and gathering information to shortlist them, soliciting price proposals from potential subcontractors and vendors, and checking references on a subcontractor qualification-based procurement process. During contracting, project engineers are often asked to review subcontracts prepared by the project manager for the scope items they will oversee the performance of. More experienced project engineers may be asked to prepare draft subcontracts for review by the project manager or the superintendent.

Summary

The project delivery method is the owner's plan for how to deliver their project. Procurement and contracting define critical operating procedures for implementing that plan. Project delivery begins with the identification of the project objectives, and it continues with the selection of a project delivery method that will help to fulfill these objectives. However, most project owners rely on other project participants for specific work items. Procuring and contracting these participants to the project team is similar to recruiting players for the football team of a university. Once the team is assembled, the project will be executed with each team member playing a role on the field that is dictated by their contract. If the wrong party is selected for a role or that party is provided with improper instructions through its contract, it may make it difficult for all parties to succeed on the project.

Review questions

1. Which procurement method should be used to place the most emphasis on price?
2. Which procurement method should be used to place the least emphasis on price?
3. Which procurement methods allow an owner to use multiple decision criteria?
4. What is a schedule of values? Can a schedule of values be used under a cost plus a fee contract?
5. Under a cost plus a fee with GMP contract, who is initially responsible for price escalation beyond the GMP? Who retains savings under the GMP?
6. Under a lump–sum contract, who is initially responsible for price escalation beyond the lump sum? Who retains savings under the lump sum?

Exercises

1. Review the information provided in the example of A+B bidding in Table 5.1.
   a. Which contractor would be awarded the contract if road user costs were estimated to be $9,000/day?
   b. Which contractor would be awarded the contract if road user costs were estimated to be $5,000/day?
2. Review the information provided in the example of a unit price contract in Table 5.2.
   a. If the actual quantities for Contract 1 were 1,300 cy, what would be the contract final price?
   b. If the actual quantities for Contract 1 were 1,600 cy and the extra quantities were discounted by 10% of the unit bid price, what would be the contract final price?

Reference

General Service Administration (GSA). (2017, January 19). Federal Acquisition Regulations (FAR): Subpart 16.4—Incentive Contracts. Retrieved September 26, 2017, from www.acquisition.gov/far/html/Subpart%2016_4.html

6 Introduction to estimating, scheduling, and project controls

Introduction

Estimating, scheduling, and project controls are the major topics in construction project management that are introduced briefly in this chapter, the last chapter in Section A, before we transition to the individual operations performed by field and project engineers. However, these topics will be discussed in detail throughout numerous other chapters, including:

Chapter 7: Safety control and reporting
Chapter 8: Production control and reporting
Chapter 9: Quality control and reporting
Chapter 10: Design review
Chapter 11: Project documentation
Chapter 12: Cost engineering
Chapter 16: Cost and schedule updates
Chapter 17: Cost estimating
Chapter 18: Planning and scheduling

In each of these chapters, and others, we will dive deeper into the field and office roles of project engineers, with each chapter progressively adding more advanced topics and responsibilities. This chapter is not a duplication of what is to come but rather just a brief introduction to these topics and the subsequent chapters.

Once a construction firm is selected for a project, various phases ensue sequentially, as shown in Figure 2.3:

- Pre-construction, including at a minimum, planning and start-up,
- Control and administration,
- Construction execution, and
- Close-out.

The estimate and schedule are developed during the pre-construction phase. Project control systems are set up during the start-up phase and occur primarily in the control phase, but they are documented all the way through the close-out phase. The natural order of progression is as shown in Figure 6.1.

Estimating

In Chapter 17, we discuss estimate types, basic estimating processes, and estimating skills. The three most common types of estimates include budget estimates, which are performed to identify early budget expectations for the project owner; bid estimates, which include lump sum and heavy-civil unit-price bids and are used to participate in competitive bidding; and guaranteed maximum price (GMP) proposals, which are popular on many privately negotiated projects.

Cost estimating is the process of collecting, analyzing, and summarizing data to prepare an educated projection of the anticipated cost of a project. Costs can be classified into *direct costs*, *indirect costs*, and *markups*. Direct costs directly relate to specific work items. For instance, the cost of building a concrete column—including self-performed labor and materials—would be a direct cost. Similarly, the cost of installing lighting fixtures by an electrical subcontractor—including labor and materials—would be a direct cost. Indirect costs refer to either jobsite overhead or home-office overhead. *Jobsite overhead* costs are the additional costs of doing construction, which do not directly relate to any specific scope of work but apply to the project as a whole. Examples of jobsite overhead costs include the salaries of the project manager, the superintendent, and the project engineers. *Home-office overhead* costs are the costs for doing business and do not relate to any specific construction project. Examples of home-office overhead costs include the salaries of individuals in the home office and the costs of maintaining the home office. *Markups* are not actual costs. Instead, they refer to the reward of doing business in the form of profit and the risk of adverse circumstances in the form of insurance and contingencies. Contract terms will affect the need for a contractor to incorporate contingencies into their price.

Similarly to the construction of a building, estimating is a logical process consisting of a series of steps. The estimating process applies to all construction firms, including prime and subcontractors. However, the role of prime contractors is broader in nature because they are responsible for acquiring subcontractors and providing a combined price proposal to the project owner.

Once upper-management decides the project is going to be pursued and is "estimate-worthy," a lead estimator is assigned. Project cost estimates may be prepared either by the project manager and project engineers or by staff estimators from the estimating department of the construction firm. All of the steps necessary to prepare a complete bid or proposal are shown later in Figure 17.1. We will briefly introduce a few of the steps here.

Quantity take-off

The quantity take-off (QTO) step involves counting work items and measuring volumes leading to the development of a bill of quantities documented in QTO sheets. An example of a QTO sheet

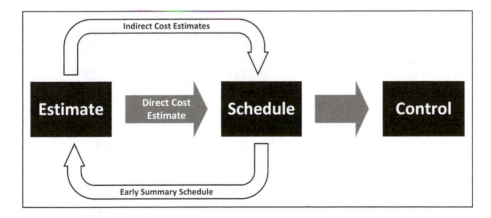

Figure 6.1 Project Controls

is shown later in Figure 17.3. The drawings are marked up to indicate that quantities have been "taken-off" the drawings and placed on the QTO sheets. The QTO process may be manually or electronically performed. A couple of straightforward examples follow.

Building project QTO example

As a junior estimator, you have been assigned the task of the door, frame, and hardware QTO. You may simply go to the door and hardware schedule and take counts, separated by type (wood doors versus hollow metal doors, hollow metal door frames versus aluminum frames), and develop totals for each one. However, the floor plans may have slight variations, so a thorough estimator would compare the door and hardware schedule against the floor plans. These quantities are then noted on the QTO sheets; they will be summarized and later transferred to pricing recap sheets. The door, frame, and hardware count may result in the following list:

- 3–0 × 7–0 single hollow metal door frames: <u>33 frames</u>
- Hollow metal door leafs: <u>42 leafs</u>
- 3–0 × 7–0 solid-core cherrywood door frames and leafs: <u>295 each</u>
- Wood slider doors: <u>160 leafs</u>
- Bi-fold doors: <u>18 each</u>
- Door hardware, one set for each leaf: <u>515 sets</u>

After successfully sharing your door, frame, and hardware QTO with the lead estimator or project manager you are working for, you may be asked to go back and look at the drawings to add counts for door signage, glass relites, and grouted hollow metal door frames.

Heavy civil QTO example

You are a project engineer for the Gateway Construction company, which is bidding on the Montana I-90 overpass project in three weeks. You have been assigned the task of estimating the volume of concrete, square feet of formwork, and number of anchor bolts needed for the F10 pile caps. You note from the foundation plan and pile cap schedule that there are 15 F10 pile caps, each of them measuring 10-feet square by 2.5-feet deep with eight embedded 1-inch diameter by 12-inch long anchor bolts (ABs), which will later receive 6-inch square hollow structural section (HSS) columns. You will note all of this on the QTO sheet, similarly to the following list:

- 15 EA @ 10' × 10' × 2.5' deep = 3750 CF (rounded)
- 3750 CF + 5% waste = 3956 CF, /27 CF/CY = <u>147 CY of concrete</u>
- 15 EA @ 8 ABs/EA = <u>120 EA 1" × 12" ABs</u>
- 15 EA @ 4 sides of 10' long by 2.5' deep = <u>1,500 SFCA of formwork</u>

After successfully showing this QTO to your project manager, you may be asked to perform the same process for the other types of pile caps and perhaps the concrete grade beams as well. The quantities are extended, totaled, and underlined, ready to be transferred to the pricing recap sheets.

Pricing

The pricing step is divided into self-performed labor pricing, material pricing, and subcontract pricing. The direct labor cost is computed by using productivity rates in man-hours per unit of work performed (i.e., MH/EA) and current local labor wage rates (i.e., $/MH). Material and subcontract prices are developed most accurately by using supplier and subcontractor quotations. Refer to Chapter 17 and the companion website for a complete pricing recap sheet example. Some simple pricing short notations from both of our examples above follow:

- Building project example: 515 door leafs and hardware sets @ 6 MH/ EA = 3,090 MHs @ $29/ MH = $89,610 labor cost (plus burden added "below the line")
- Heavy civil project example: 120 EA 1" × 12" ABs @ $18.10/ EA to purchase = $2,172 material cost
- Heavy civil project example: 147 CY of concrete @ $12.50/ CY to pump = $1,838 pumping subcontractor

Note: In an effort to use gender-neutral language to reflect an increasingly diverse workforce, some companies are replacing the term man-hours with person-hours (PH). This terminology is less common and not widespread, so we decided to follow the traditional terminology in the rest of this book.

Summary and markups

The estimate summary page is divided into "above-the-line" and "below-the-line" sections. Costs calculated on the pricing recap sheets and transferred to an estimate summary page are considered as direct costs; they are included *above the line* on the estimate summary. Jobsite overhead costs (also known as general conditions) are also construction costs but are considered as "indirect" costs, in that they are not directly related to any specific scope of work but rather apply to the whole job. Development of a general conditions estimate is schedule-dependent. If the project is 14 months (i.e., 60 weeks) long and we are going to employ a superintendent at a wage of $2,200 per week, then we need to include $132,000 in our general conditions for the superintendent. Even though general conditions are considered as indirect costs, they are usually included *above the line* and regarded as a cost of the work, but contract terms have some bearing here.

There are a variety of percentage add-ons or markups *below the line*. Four of the most common markups include labor burden—which includes fringe benefits plus labor taxes (if not included with the direct and indirect wage rates on the pricing recaps)—liability insurance, excise tax (state-dependent), and fee. Home-office indirect costs are combined with desired profit to produce the fee markup. The final fee and the final total bid at the bottom of the estimate summary page are often determined by the contractor's officer-in-charge (OIC) at the last minute before the price is turned in to the client. The fee calculation depends upon several conditions including company volume, market conditions, labor risk, and resource allocations. Jobsite general conditions and summary estimates for our case studies are included in Chapter 17, and the companion website hosts more detailed versions.

Scheduling

Many see planning and scheduling as the same operation, but they are slightly different, with proper planning preceding development of the project schedule. Project engineers will not have as much of a role in planning as they will in scheduling but should welcome the opportunity to become involved in development of the project plan whenever it is offered. Planning is usually performed by the project manager and superintendent with the assistance of other experienced and specialized contractor personnel. Planning includes performing the following tasks:

- Developing a *work breakdown structure* (WBS);
- Identifying a logic for work activities and restraints on starting or completing activities;
- Evaluating the availability of manpower for self-performed work;
- Developing a subcontracting plan, including what work is to be subcontracted versus self-performed;
- Estimating activity durations from the labor hours on the direct-work pricing recap sheets and from subcontractor input;
- Forecasting material and equipment delivery dates;
- Identifying owner/architect restraints, such as design package releases, receipt of permits, and delivery of owner-supplied equipment; and
- Selecting means and methods of construction, including choices of concrete formwork (rented versus site-fabricated), internally or externally rented equipment, and hoisting (tower crane or crawler and/or forklift).

The development of the actual schedule document is a more simple and mechanical process than the development of a proper project plan. It is not as simple as entering all of the above information into scheduling software and plotting a schedule, although some do it that way. A good detailed contract schedule will go through several rough drafts and edits, each of which is reviewed by the superintendent and his or her foremen, along with key subcontractors.

Although we said at the outset of this chapter that the estimate leads to the schedule and both lead to project controls, an early summary schedule is actually necessary to complete the jobsite general conditions estimate and to verify if the planned approach would meet any contractual deadline or milestone. However, an accurate summary schedule will not be developed until after the detailed construction schedule is completed. Summary schedules may be as short as 15–20 line items or up to 40–50 line items long. Summary schedules may also be included with responses to requests for proposals (RFPs), and they are sometimes attached to the prime construction contract agreement as an exhibit.

Many project owners and contractors prefer a detailed schedule to be attached to the contract as an exhibit. Detailed construction schedules are often over 100 line items long, with many stretching to thousands of line items on larger complex projects, especially lump sum public works projects. The schedule is a construction management tool and should follow the *Pareto principle*, also known as the *80–20 rule*, in that 20% of the activities will require 80% of the time. The schedule should be detailed enough that progress can be accurately measured and foremen and subcontractors can develop their three-week look-ahead schedules from it. However, it should not be so detailed that management of the schedule takes on a life of its own. For instance, the auger-cast piling should be shown, along with the pile caps and the formwork for the pile caps, but it is not necessary to treat the application of spray-on grade beam form oil as a separate item. The detailed schedule should be posted on the meeting room walls in the general contractor's jobsite trailer, and the superintendent should provide the project owner with a schedule status update each week in a meeting. For building projects, this meeting is usually called the owner–architect–contractor (OAC) meeting. A summary schedule and a three-week look-ahead schedule are included in Chapter 18, and a more detailed contract schedule is included on the companion website.

Project controls

In Chapter 2, the task of evaluating the success of built environment projects was introduced, together with the four measurable dimensions that constitute the pillars to success of built environment (BE) projects: *cost*, *time*, *quality*, and *safety*, as shown in Figure 2.1. Project controls include a set of processes put in place for evaluating the achievement of project success throughout the project delivery. Thus, project controls adopt processes to evaluate the achievement of any of the four stated objectives. This approach has progressed from the historical approach of assessing the achievement of cost, schedule, and quality by adding a fourth set of processes designed to evaluate safety goals, which is of foremost importance in the construction industry. Moreover, Figure 2.1 illustrates the importance of setting up a document control system according to contract requirements, so that information is collected and is easy to retrieve for controlling the project. Whereas document control does not directly measure the achievement of project success, a sound approach to document control contributes to success on a project. Therefore, this textbook will also discuss the main document control processes. In this section, we will briefly introduce each of these controls, and they will all be elaborated on more fully in subsequent chapters.

Can we really "control" our craftsmen and subcontractors? Can we force them to do something for a given cost and in a set amount of time and at the level of quality that we require, all while staying safe? What if they don't do it? Can we penalize them? We can terminate a direct-hire employee who does not perform quality work. We can require a subcontractor to remove their employee who violates safety rules or even terminate a subcontractor that is not meeting the schedule, although this is more difficult than it sounds. Our way to best "control" is actually to actively "manage" the process well. Project controls are achieved by having the right people in the right place at the right time and providing them with the necessary tools and equipment so that they can do their best to match cost, schedule, quality, and safety expectations and document everything according to contract requirements.

Document control

The control of documents does not guarantee or directly measure project success. Therefore, document control is not considered a dimension of success to be measured and controlled over time. However, the setting up of a document control system that is appropriate for the given project is one of the most important tasks for the project manager and project engineer during project start-up. This system will allow the project team to accurately track down and document success in safety, cost, time, and quality.

Safety control

Whereas safety was only recently added to the dimensions for measuring construction project success, most construction firms now predicate the concept of "safety first" in their core business practices. Safety control is accomplished similarly to quality control, with clear expectations and continual open communication. Even if a project has a full-time safety inspector or the corporate safety officer visits the site weekly, most see the general contractor's (GC) superintendent as responsible for safety control—this is similar to expectations of quality control. General contractors and their superintendents have been found in some situations to be not only contractually responsible for safety, but legally responsible. Regardless of who is officially designated as the point person for safety control, we are all responsible for observing potential safety hazards and reporting any violations. Some of the proactive safety control efforts (further described in Chapter 7) performed by the superintendent, with assistance from the project engineer, include:

- Preparing a project-specific safety plan for submission to the project owner,
- Employing only subcontractors with proven safety records,
- Requesting subcontractors to sign up to the GC plan and/or to prepare project-specific safety plans for submission to the GC for approval,
- Conducting new subcontractor and craftsmen safety orientation meetings, and
- Holding weekly safety meetings, usually first thing Monday morning, and requiring attendance for all on site.

If safety violations occur, the superintendent can follow different pathways, which are listed in order of severity: (a) issue a written warning, (b) terminate a direct or subcontractor craftsman, (c) fine or back charge a subcontractor, or (d) terminate the subcontractor. The severity of the violation will suggest to the superintendent which pathway is the most appropriate. If accidents occur and someone is injured, there are additional inspections and documentation required, as further discussed in Chapter 7.

Cost control

Cost control and schedule control are both elaborated on in Chapter 8. There are five phases of cost control. The first phase begins with an accurate estimate and schedule. All five of our cost control phases are depicted in Figure 6.2. This same figure will also be discussed later in Chapters 8 and 16. Many of our chapters will connect to these phases in more detail, along with the role of the project engineer, but what follows here is a brief introduction to each phase.

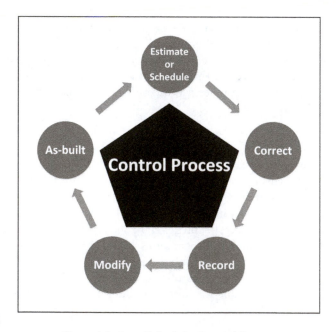

Figure 6.2 Cost/Schedule Control Process

Estimate

As introduced earlier, the estimate is an assembly of measured material quantities, competitive market-rate material unit prices, historical direct labor productivity rates, current direct labor wage rates, competitive subcontract quotes, and a series of markups and fees. To better prepare the management team for cost control, the estimate should be assembled by work packages and every line item in the estimate assigned an individual cost code.

Correct

No estimate or schedule is perfect. After the contractor receives notice of award of its bid or proposal, the estimate must be corrected with the actual subcontract and purchase order buyout values (see Chapter 19) and entered into the company's cost control system. If the estimator made any mistakes, these must be corrected now, with potential modifications to contingency funds or fees. The superintendent and project engineer cannot begin construction and effectively control costs with an incorrect estimate.

Record

This is the largest and most time-consuming phase in the cost control process. The recording of actual costs is also a function most often performed by project engineers. Some of the ways the engineer is involved in recording costs include:

- Assisting foremen and superintendents with development of work packages,
- Entering direct labor cost codes on time sheets,
- Entering cost codes on short- and long-form purchase orders and subcontract agreements (see Chapter 19),
- Verifying that supplier and subcontractor monthly invoice amounts match their contract values,
- Entering cost codes on supplier and subcontractor invoices, and
- Assisting the project manager with monthly fee forecast development (see Chapter 12).

One of the book's authors spent an entire year, including a very cold winter in Pennsylvania in a triple-wide trailer, manually entering cost codes daily for 1,000 pipefitters on a power-plant project and then researching incorrect entries the following morning. Project engineers have to pay their dues in order to progress into project management, and it sometimes starts with cost coding, so this should be looked at as a project engineering "get-to" and not a "have-to."

Modify

As stated, there is not a perfect estimate nor will many construction projects proceed "exactly" according to the original plan and schedule. Continual comparison of actual costs recorded against the estimate will uncover variances that will warrant attention and potential adjustment by the project team. The contractor cannot wait until the project is finished to find out if it has made money. At that time, there is nothing that can be done to fix the problem. Costs may differ from the estimate for a variety of reasons, including:

- The estimate was plagued by errors,
- The project team is not implementing construction according to the original plan,
- Subcontractors or suppliers are not performing,
- Self-performing field crews are not performing,
- Site management team members, including the project manager (PM) and superintendent, are not performing,

- The GC is performing work that was the responsibility of a subcontractor or the project owner,
- The project owner and/or designer have added scope or affected the planned work sequence,
- The occurrence of inclement weather,
- The discovery of unforeseen conditions,
- The schedule has been lengthened, or
- There is an abuse of overtime.

Once the reason for the cost overrun is understood, the plan or process may need to be adjusted or modified by the project team; with luck, the cause will be discovered soon enough to implement the adjustment. Some of the corrective actions or "fixes" might include:

- Enforcing subcontract agreements,
- Back charging subcontractors,
- Changing subcontractors (this is difficult to do),
- Submitting a change order proposal to the project owner (see Chapter 20),
- Changing in-house GC personnel,
- Utilizing different means and methods or equipment,
- Expediting the schedule, or
- Using overtime selectively (sometimes spending money can save money).

As-built

An as-built estimate, as well as an as-built schedule, is an important historical tool that experienced contractors utilize to successfully procure future construction projects. The time to start development of the as-built estimate is when the cost control process is started, with a detailed and cost-coded estimate. During the course of construction, material costs and labor hours are accurately recorded, along with measurement of the actual quantities installed, to develop as-built material unit prices and as-built labor productivity rates. The as-built estimate is definitely part of the close-out process (see Chapter 14), but the sooner it is prepared, the more accurate the results will be. These figures should be shared by the project team with staff estimators for incorporation into the contractor's estimating database. In this manner, the last phase of cost control, the as-built phase, is actually the first phase for the next project, and the cost control cycle in Figure 6.2 repeats itself.

Time control

The task of updating and modifying costs and schedules is discussed in Chapter 16. As indicated earlier, we do not necessarily "control" schedules or time as much as we manage the process. This begins with a detailed and collaboratively developed schedule, not a top-down schedule. A discussion on the importance of a collaborative scheduling process is provided in Chapter 18. Project engineers can be tasked to assist the superintendent with recording actual schedule statistics. Marking up of the contract schedule with actual start and completion dates, along with actual material delivery dates, is integral to development of the as-built schedule. The project superintendent will provide a status report on the schedule at the weekly project team meeting. Three-week schedules developed by the GC are also shared with the project owner and designer during this meeting. Subcontractors will provide the superintendent with their own three-week look-ahead schedules, often at the Monday morning foremen's meeting. As will be discussed in Chapter 16, a schedule only needs to be revised and reissued if there are major deviations. New schedules must be incorporated into the prime contract and each subcontract and potentially lead to cost ramifications.

Quality control

Before any attempt is made to control quality, an understanding of quality is foremost. The concept of quality is fluid and subjective. Where users may judge a project's quality based on its ability to fulfill the user's needs, the project owner may additionally judge a project's quality based on its "bang for the buck" or its ability to achieve excellence. This will vary widely, depending on whether the project is for the project owner's headquarters or for a suburban warehouse. In the absence of clear indications, a construction firm would judge quality based on the project conformance, as described in the contractual documents. Therefore, it is important that the project owner conveys their quality expectations through these documents.

However, quality control (QC) management first starts with the project owner employing the most qualified design professionals and giving them adequate time and financial resources to prepare a quality design. The next step is for the project owner and designer to clearly communicate their well-defined and reasonable quality expectations in the contract documents. This should all be discussed at the owner–designer–contractor pre-construction meeting. Just as the project owner needs to select a quality GC, the GC needs to select qualified subcontractors and suppliers. Similarly to the design team, all of these firms need to have adequate cost and time allowances, proper materials and equipment, and most importantly, qualified field craftsmen in order to do a quality job. As discussed more fully in Chapter 9, the superintendent and project engineer manage quality among the subcontractors on the jobsite in a variety of fashions, including:

- Pre-construction meetings,
- Submittal approvals,
- Thorough and timely requests for information (RFIs),
- In-process work inspections,
- Follow-through with non-conformance reports (NCRs) and QC log maintenance, and
- Final punch lists and work acceptance.

Project engineering applications

The project engineer assists the project manager and superintendent with a variety of estimating, scheduling, and project control functions. These roles and responsibilities are described in the upcoming chapters in detail. We have subdivided our textbook into increasingly advanced project engineering areas of responsibility, which may correspond into dividing the position or title of "project engineer" into loose subdivisions. Titles vary depending on the contractor and the project. Many will use the generic term "project engineer", as we will do as well. However, we will later refer to different project engineering duties as follows:

- Field engineering is intended as assisting the superintendent on the jobsite; Section B, Chapters 7–9.
- Project engineering is intended as assisting project managers, estimators, and schedulers, whether located in the jobsite or main office, job-size dependent; Section C, Chapters 10–16.
- Senior project engineering is intended as assisting the project manager in all or some of the PM functions and potentially runs smaller projects as the PM; Section D, Chapters 17–22.

Summary

The summary section for this chapter is really a brief introduction to the rest of our book. Estimating is the process of measuring and counting all of the material quantities, "taking them off" the drawings, and applying market material rates and historical labor productivity. Subcontractors perform 80% or so of the field work on a construction project, so incorporation of competitively bid subcontract prices is paramount to a successful GC estimate. The GC's officer-in-charge will input the fee and other markups that are applied "below-the-line" on the estimate summary.

Scheduling starts with good planning. There are many types of schedules used on a construction project, including summary, detailed, or contract, as well as three-week look-ahead schedules. But schedules are only useful as construction management tools if they are collaboratively developed by the GC, including its superintendent, and with input from its subcontractors. We will discuss four different project control processes in this textbook, namely, cost, schedule, quality, and safety control. Essential to the contractor delivering a successful project to a satisfied client, all success-related types of controls rely on appropriate document control, with project engineers being very involved in processing the documents for each of these processes.

This chapter was just a brief introduction and overview to many construction management and project engineering operations. Hopefully, it will have sparked interest in the reader. Forthcoming chapters will continue to build upon these concepts.

Review questions

1. What are the four types of project controls?
2. What is the difference between planning and scheduling?
3. Which occurs first, the detailed construction estimate or the detailed construction schedule, and why in that order?
4. What is the difference between "manage" and "control" with respect to any of our project management control functions?

Exercises

1. Prepare an argument for why you as a contractor would want to (a) include only the summary schedule as a contract exhibit or, conversely, (b) include a detailed schedule as the contract exhibit.
2. Prioritize the four types of controls. If you had to drop one in order to get the project completed, which would it be?
3. Prepare an argument for why you feel a recent civil engineering or construction management graduate would learn the most by being (a) in the jobsite trailer working for the superintendent as a field engineer, (b) in the home office assisting staff schedulers and estimators, (c) a project engineer assisting one project manager in all of his or her capacities in the trailer on the jobsite, or (d) a project or office engineer working out of the home office but assisting several project managers on several projects with their RFIs, submittals, and meeting notes.
4. Other than the items bulleted above under cost control/modify, list another reason costs may exceed the estimate and an alternative solution for that problem.

Section B
Field engineering

7 Safety control and reporting

Introduction

Construction is an extremely dangerous industry. According to the Occupational Safety and Health Administration (OSHA) and the Bureau of Labor Statistics (BLS), 937 fatalities occurred in construction in the U.S. during calendar year 2015, which is equal to approximately 21% of private industry employment-related fatalities during the same calendar year (BLS 2016). In addition to fatalities, construction jobsite injuries and illnesses affect the construction industry's workforce more than most others. Most construction jobsite fatalities are connected to four causes that are named by OSHA as the Construction Fatal Four and account for about 64% of construction fatalities: Falls (~39%), Struck by Object (~10%), Electrocutions (~7%), and Caught-in/between (~7%) (OSHA 2017). The latter group includes construction workers killed when caught-in or compressed by equipment or objects and workers killed when struck, caught, or crushed in collapsing structures, equipment, or materials. These statistics show the importance of safety in construction.

Fatalities, injuries, and illnesses affect the construction workforce, so there is a moral obligation for the industry as a whole to do better. However, there are many additional ramifications associated with poor safety performance in construction, including:

- Jobsite medical attention,
- Damage or destruction of materials, and cleanup and repair costs,
- Unproductive labor time, and potential loss of trained labor,
- Workforce morale,
- Construction schedule delays or work slowdown, and
- Lawsuits and other legal and administration expenses.

In addition to fatalities, injuries, and illnesses, narrowly avoided incidents (also known as near misses) are indicators of safety shortfalls because these events have the potential to result in fatalities, injuries, illnesses, or damage.

OSHA does not only compile statistics; it has the responsibility of overseeing safety standards on construction projects in addition to other industries. The Occupational Safety and Health Act established OSHA, but it allows states that desire to administer their own industrial safety programs to do so, as long as their requirements are at least as stringent as those imposed by OSHA. Many states, therefore, have their own self-administered programs.

Control of safety performance should not be limited to counting fatalities, accidents, and near misses. Instead, it should continually evaluate workers' exposure to hazards and implement measures to reduce them; safety professionals consider this to be the best approach for protecting workers. To this end, control measures have been categorized hierarchically to help implement

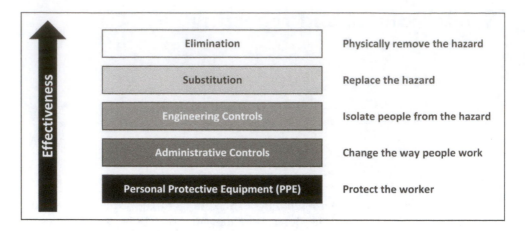

Figure 7.1 Hierarchy of Hazard Controls
(Adapted from NIOSH 2016)

comprehensive and effective safety control programs. Figure 7.1 shows the *Hierarchy of Controls* as promoted by the National Institute of Occupational Safety and Health (NIOSH).

During the initial planning phases, construction firms can pursue elimination or substitution of hazards through their selection of construction processes, including their choice of means and methods. Elimination and substitution of hazards are considered the most effective safety control measures. However, once construction processes have been selected, elimination and substitution are difficult—if not infeasible—to implement, because each construction process carries its own hazards. However, safety knowledge and technology provide opportunities for implementing engineering controls that will isolate people from these hazards. For instance, dust exposure is a known hazard for drywall craftsmen. This exposure is heightened during drywall sanding activities. Nonetheless, drywall sanding is often necessary to achieve the project objectives. Thus, if the dust exposure hazard exists, how can we isolate our drywall craftsmen from it? NIOSH research has shown that use of a pole-sanding vacuum control system is expected to substantially reduce dust exposure from hand-sanding. Therefore, if these craftsmen are provided with the right tools, it will reduce their exposure. The next levels of hazard control include changing the way people work through administrative measures or protecting workers with personal protective equipment (PPE). These measures are widely discussed throughout the rest of this chapter.

At the industry level, construction firms are encouraged to improve safety through a carrot-and-stick approach. Construction firms that have a poor safety performance will be subject to higher fines by OSHA or state agencies, which will increase insurance rates, a somber workforce morale, and a bad reputation among other construction firms and project owners. On the other hand, construction firms that are able to improve their safety performance will improve their reputation and workforce morale while reducing or nullifying fines and decreasing insurance rates.

Construction firms operate their business through projects, so improvements to safety at the project level are crucial. There are three major aspects of site safety: (a) safety of those working on the site, (b) safety of those visiting the site, and (c) safety of the general public near the project site. Safe operations of construction equipment relate to all three aspects. Construction firms produce project-specific safety plans to address all of these aspects.

Safety is a complex issue, and this chapter only provides an overview on a few safety control and reporting topics, including:

- Developing project-specific safety plans and job hazard analyses (JHAs);
- Implementing industry-wide safety control measures, including workers' compensation insurance programs, drug and alcohol abuse monitoring, and communications systems for hazardous materials;
- Implementing project-specific safety control measures, including continuous supervision, crew-level pre-task safety checklists, weekly safety meetings and inspections, and safety training; and
- Reporting safety issues, including inspection reports, and near misses and accident investigations.

Project-specific safety planning

Most large building and civil construction companies have an in-house company safety program for their direct employees and their projects. The effectiveness of these programs, however, is directly related to the executive management's commitment to safety. In addition, project-specific safety programs are developed for each project. The site management, including project managers and superintendents, are responsible for the safety of workers, equipment, materials, and the general public on their project sites. Their responsibility is not only ethical but often contractual and even legal. The site management must set the standard regarding safety on their projects, lead workers by example, and enforce safety procedures at all times throughout the project. To this end, a continual safety awareness campaign is required, which focuses on reducing hazards. Standard elements of project-specific safety programs include a project-specific safety plan—which relies on comprehensive and task-specific JHAs—and ongoing safety initiatives designed to control hazards.

Project-specific safety plan

Potential safety hazards exist on all construction projects, and project-specific safety plans are necessary to proactively manage these hazards. Project-specific safety planning is a logical process that examines the potential hazards associated with construction activities. The planning process requires an understanding of construction practices and the site-specific conditions in which they will be performed. Public work bid projects require the general contractor to submit such plans at pre-construction conferences. Private and negotiated clients may require that these plans be submitted with the proposal package. A project-specific safety plan serves many purposes, such as:

- Identifying key personnel and their safety roles and responsibilities,
- Establishing the chain of communication on safety,
- Describing specific safety awareness training and orientation programs to be utilized,
- Listing the hazard identification and control methods for the prevention of incidents,
- Identifying an emergency response plan, and
- Defining the inspection requirements to evaluate the project's safety program.

Comprehensive and task-specific job hazard analyses

Construction jobsite accident prevention starts with a safety hazard analysis of each construction task that is to be performed. Therefore, initial steps for implementing a project-specific safety plan

Gateway Construction Company
2201 First Avenue
Spokane, WA 99205
509-642-2322

JOB HAZARD ANALYSIS

Project: I-90 Overpass, Montana, Job 1732
Phase: 1: Structural Excavation, Shoring, and Foundations

Construction Activities	Potential Hazards	Hazard Mitigation
1. Excavation	Equipment striking workers	Morning safety meeting Bright clothing Backup alarms
	Fall into excavation	Guardrail Bright flagging
2. Auger cast piling	Equipment striking workers	Morning safety meeting Bright clothing Backup alarms
	Slipping on waste and spills	Backhoe cleaning waste Dedicate one laborer
3. Vertical rebar	Impalement	Orange caps on rebar
4. Flagger	Struck by traffic	Certified flaggers only Proper clothing and signs Coordinate with police
	Struck by our equipment	Morning safety meetings Certified flaggers only Proper clothing and signs
5. Material deliveries	Truckers striking workers	Stop trucks at gate Inform of safety rules
	Truckers striking equipment	Stop trucks at gate Inform of safety rules

Prepared by: Randy Buckwater, Superintendent
CC: Client, Designer, Subcontractors, PE, Foremen

Figure 7.2 Comprehensive Job Hazard Analysis

include performing an initial comprehensive identification and analysis of potential hazards that will guide the identification of task-specific hazard mitigation measures. An example of a comprehensive hazard identification and analysis is shown in Figure 7.2. Comprehensive JHAs are often followed up by task-specific hazard analyses, such as the one shown in Figure 7.3 for the glazing subcontractor on a tenant improvement project.

The comprehensive JHA should be prepared during the project planning process. This JHA becomes the basis for the project-specific safety plan and usually includes:

- A description of the construction processes,
- A list of the hazards associated with each process, and
- The plan for addressing each hazardous situation.

Task-Specific Job Hazard Analysis (JHA)

Activity/Work Task: Install glass curtain wall and wall panels	Overall Risk Assessment Code (RAC) (Use highest code)	**M**

Project: Tenant Improvement of 2nd Floor Restaurant Lounge

Risk Assessment Code (RAC) Matrix

Contractor: Old Fashioned Glazed Visions (OFGV)

Date Prepared: 3/9/17	Severity	Probability				
		Frequent	Likely	Occasional	Seldom	Unlikely
Prepared by (Name/Title): Jill Smith - OFGV Site Superintendent	Catastrophic	E	E	H	H	M
	Critical	E	H	H	M	L
Reviewed by (Name/Title): Jill Smith - OFGV Site Superintendent	Marginal	H	M	M	L	L
	Negligible	M	L	L	L	L

Notes: (Field Notes, Review Comments, etc.)

Step 1: Review each "Hazard" with identified safety "Controls" and determine RAC (See above).

"Probability" is the likelihood to cause an incident, near miss, or accident and identified as: Frequent, Likely, Occasional, Seldom or Unlikely

Competent Person(s):
Jill Smith - OFGV Site Superintendent
Jack Spade – OFGV Glazier Foreman

"Severity" is the outcome/degree if an incident, near miss, or accident did occur and identified as: Catastrophic, Critical, Marginal, or Negligible

RAC Chart

E = Extremely High Risk	
H = High Risk	
M = Moderate Risk	
L = Low Risk	

Step 2: Identify the RAC (Probability/Severity) as E, H, M, or L for each "Hazard" on AHA. Annotate the overall highest RAC at the top of AHA.

Job Steps	Hazards	Controls		RAC
1. Lifting glass up to the lounge	1. Sharp corners on the frame, smashing fingers while drilling holes in frame and running IGU cables	1. Gloves 100%, keep work area clean and keep fingers away from area where drilling is done.		L
2. Stage glass	2. Fall hazard	2. Keep hands clear or use a block when installing stops and snap cover.		L
3. Remove old glass	3. Trip/slip	3. 100% tie off when removing and installing lower glass. Danger tape work area.		M
4. Drill hole for IGU cable	4. F.O.D.	4. Cleans you install all, thrash of paper must go into a garbage can or bag.		L
5. Install new glass	5. Protecting existing frame	5. Handle material with caution, protect from scratches and damage.		L
6. Hook up IGU cable	6. Pinch points	6. 100% eye protections		L
7. Install stops and vinyl	7. Particle in eye	7. Inspect load before forklifting.		L

Equipment to be Used	Training Requirements/Competent or Qualified Personnel name(s)	Inspection Requirements
● Reach forklift ● Drill, impact guns and hand tools	● Forklift card ● None	● Inspect forklift daily ● Quarterly inspection and labeling of electric power cords

Figure 7.3 Task-Specific Job Hazard Analysis

Once subcontractors are secured to the project, task-specific JHAs can be requested from them or produced as part of work planning exercises for self-performed activities. Comprehensive and task-specific JHAs are among the construction management tools utilized in active safety control programs.

Safety control

Several safety initiatives designed to control hazards constitute another element of project-specific safety programs. Some of these initiatives are implemented industry-wide, such as workers' compensation insurance programs, drug and alcohol abuse monitoring, and communications systems for hazardous materials. Other initiatives are implemented differently across contractors and projects, including continuous supervision, crew-level pre-task safety checklists, weekly safety meetings and inspections, and safety training.

Workers' compensation insurance programs

State workers' compensation laws mandate coverage for all personnel on a jobsite. Workers' compensation insurance is a no-fault insurance program that protects a contractor from being sued by their employees as a consequence of injuries sustained on the jobsite and provides compensation to workers who are injured on the job, regardless of who is at fault. Workers' compensation benefits cover:

- Medical expenses,
- Lost wages when unable to work,
- Rehabilitation for a new vocation if necessary, and
- Pensions for long-term disabilities, if necessary.

Many states control workers' compensation insurance and provide access to this insurance through state-administered funds, whereas others rely on private insurance companies. Workers' compensation premiums have two components. The first is the base premium that is applied to each separate work classification; different crafts have different rates based upon their frequency and severity of accidents. The second component is the premium modifier, known as the experience modification rating (EMR). This is based on the volume of claims made by a contractor's employees. Typical company rates range from 0.5 to 2.0, with 1.0 serving as the baseline. Companies with good safety records typically have EMRs below 1.0. Since a contractor must pay workers' compensation premiums for each hour that each craftsman is on the jobsite, this significantly adds to the labor cost. Firms that have poor safety records have high EMRs, which results in labor costs that are higher than for those with good safety records and low EMRs. As previously discussed, this is one of the carrot-and-stick approaches that regulate safety at the industry level. Therefore, project managers, project engineers, and superintendents are incentivized to emphasize safety on their projects, so their employers can earn low EMRs and have competitive labor rates, which would place them in a position to be able to post lower bids and win more construction projects.

To reduce the volume of claims, injured construction craftsmen should be returned to the jobsite as soon as is reasonable. Any craft employee injured on a project needs to be taken to a physician for examination and determination of the extent of the injuries and the type of work, if any, that the injured worker is able to perform. Based on the physician's instructions, the construction company should devise a return-to-work strategy for the craftsman. This may include recuperation, physical therapy, reduced work hours, and/or modified work assignments. Return-to-work programs should be designed to benefit the injured worker and reduce the workers' compensation claims cost, which will affect the company's EMR.

Drug and alcohol abuse monitoring

The construction industry is one of the most dangerous in the United States. Craftsmen under the influence of drugs or alcohol on the jobsite pose serious safety and health risks to themselves and all those who work on the site. To combat drug and alcohol abuse, most contractors have established policies prohibiting substance use prior to or while on the site. A key element of this program is a requirement for periodic drug and alcohol testing. The testing requirements usually apply to all individuals working on the project site, whether they are employed by the contractor or by a subcontractor. Subcontractors generally are required to administer their own substance-testing programs, or they are not allowed to work on that project. Individuals refusing testing or failing to pass a test are removed from the jobsite. Many contractors require testing during the hiring process, after an accident, or when a foreman or superintendent believes an individual's behavior warrants testing. Some states allow random testing, but there are different rules on enforcement. It is a good practice for contractors to use third-party testing companies to ensure fairness in implementation and compliance with legal requirements. Selected administrators should stay up to date on the rules and court cases at state and national levels.

Hazardous materials communications

OSHA requires chemical manufacturers, distributors, and importers to provide a safety data sheet (SDS)—formerly known as a material safety data sheet (MSDS)—to communicate hazards for all chemical products. SDSs are short technical reports that identify all known hazards associated with particular materials and provide procedures for using, handling, and storing the materials safely. They also include information on first aid measures. SDSs must be provided for each material used on site. A copy of the SDS for each hazardous chemical must be made available to all craftsmen before a material is utilized on the project site. The project engineer usually collects these as a required submittal from all suppliers and subcontractors. Copies of SDSs for all construction materials should be maintained in an accessible on-site location.

Ongoing project-specific safety activities

Most construction projects are unique, and construction workers are constantly expected to familiarize themselves with new working environments. However, craft workers may only work on a project site during certain phases of work and then move on to another project. This ongoing change in the composition of the workforce presents significant challenges to the project team. Therefore, once the construction activities begin, continual safety awareness is maintained through continuous supervision, crew-level pre-task safety checklist exercises, mandatory training orientations, and weekly safety meetings and inspections.

Continuous supervision

Accidents usually occur on construction sites due to a lack of safety procedures and/or implementation of those procedures. Accident prevention requires a commitment to safety, proper equipment and construction procedures, regular jobsite inspections, and good planning. Everyone on the jobsite must understand the need to work safely and be alert for any unsafe conditions. The general contractor's (GC's) superintendents should:

- Enforce proper use of PPE by craftsmen,
- Require PPE be worn by all on the site,

- Coordinate efforts to keep the jobsite clean, and
- Teach emergency procedures to all craftsmen.

Safety procedures and techniques are identified for each element of work to minimize the potential for accidents. Many methods are utilized to reduce the risks associated with safety hazards, including:

- Modifying construction techniques,
- Keeping employees away from the hazard,
- Displaying warning signs, including backup alarms on equipment,
- Training, and
- Requiring PPE.

As shown in Figure 7.1, PPE is the last resource for controlling hazards. Therefore, when it is prescribed, each jobsite employee must wear the appropriate PPE to mitigate loss. The jobsite management team identifies general PPE requirements as part of the job hazard analysis and requires that all contractor and subcontractor craftsmen wear it. These requirements are specified in the accident prevention section of the project-specific safety plan. Usually, the minimum requirements include wearing a hard hat, safety glasses, and safety shoes at all times when working on the site. However, some contractors may be more restrictive as part of their company safety programs or because of specific hazards on the given projects. Similarly, some tasks may require additional PPE. Several types of PPE are available, but some of the most commonly encountered on a construction site include:

- Hard hats,
- Tight-fitting, bright protective clothing,
- Safety glasses or goggles,
- Safety shoes and gloves,
- Earmuffs or ear plugs,
- Personal fall protection systems, such as safety harnesses, and
- Respirators.

Crew-level pre-task safety checklist exercises

Project-specific safety plans frequently prescribe pre-task safety checklist exercises at the beginning of each work shift. These exercises are conducted by each crew to identify daily tasks and evaluate corresponding hazards and control measures. They are a proactive, self-administered approach to remind craftsmen about safety measures through a self-guided process. Sometimes, foremen or superintendents lead these exercises and may integrate them with pep talks, which are short talks frequently based on anecdotes around the hazards that the craftsmen will be exposed as part of their daily tasks.

Training and mandatory orientations

Project-specific safety plans also prescribe training requirements. Standard training requirements include mandatory safety orientations for all workers and visitors. As part of the orientation, minimum requirements for PPE are also discussed. In addition, training is conducted for hazards associated with each element of work, and mitigation strategies are required from all work crews prior to allowing them to start work. This task is often assigned to one of the project engineers. In addition, most superintendents require safety meetings and morning stretch and flex exercises prior to allowing the workers to start their shifts. Monday morning safety meetings also include refreshers on potential risks and mitigation methods for work scheduled to be performed that week. Requirements on training certifications and the identification of competent persons are also listed in the project safety plan. Under these requirements, each subcontractor should produce certifications for their craftsmen and identify a competent person who will enforce their safety plan, including any task-specific requirement for PPE that is more restrictive than those applicable to the whole jobsite.

Weekly inspections

In addition, weekly safety inspections are usually conducted at the jobsite to identify hazards and ensure compliance with job-specific safety rules. These inspections are often performed by a home-office safety officer or an on-site project engineer on smaller projects. During these inspections, hazards are identified and categorized by their gravity, and the jobsite is evaluated against other safety requirements in the project-specific safety plan. It is common to encounter a three-tier approach for categorizing hazards:

- Class A: These hazards are the most serious because they have the potential to result in injuries, permanent disabilities, and even fatalities. They may also refer to hazards that could result in an extensive loss of structure, equipment, or material.
- Class B: These hazards also have serious consequences because they can result in recordable incidents, which would require medical aid or modified work assignments, serious injuries or illnesses, temporary disabilities, and disruptive but not extensive property damage.
- Class C: These hazards may cause minor, non-disabling injuries or illnesses or non-disruptive property damage.

This categorization system is designed to guide the inspectors in their follow-ups. Whereas communication of Class C hazards to the foremen or subcontractor may suffice, the occurrence of Class B hazards will require immediate escalation to the project manager and superintendent. Similarly, some project-specific safety plans prescribe immediate escalation of Class A hazards beyond site management.

Weekly meetings

Lastly, whereas every project meeting should address safety in some manner, most of the general contracting firms conduct weekly safety meetings on Monday mornings with all workforce on site, including direct-hire employees and subcontractor craftsmen. These meetings are used to review any major hazard that is expected for the week to come, as well as near misses and injury events that occurred in the previous week. In addition, safety personnel from the subcontractors may be invited to present on a rotating basis to share with all of the workers on site their expected hazards, how they plan to control them, and what other crews need to know to not impede their safety efforts. A significant challenge to the organization of these meetings is the increased employment of workers for whom English is a second language, which requires safety meetings to be conveyed in languages other than English.

Equipment control

Most building contractors will rent their equipment from outside sources or require each subcontractor to provide their own equipment. Conversely, heavy-civil contractors often own many pieces of equipment. Whereas equipment ownership provides opportunities for improved profits, it also increases the contractor risk exposure. All contractors must pay attention to the safety implications associated with the use of construction equipment through performance of periodic inspections. However, heavy-civil contractors carry more responsibilities, some of which include management of the following:

- Maintenance log;
- Backup alarms;
- Use of direct-hire operating engineers (OEs) versus employment of operated equipment, which must be contracted as a subcontractor, not a supplier;
- Licensed and/or certified OEs;
- Regulated breaks for OEs;
- On-ground equipment coordinators, foreman, potentially "oilers" and "riggers;"
- Communication with flaggers;
- Maintenance check for hydraulic leaks;
- Environmentally controlled refueling stations;
- On-ground bright protective clothing for all in the vicinity (if not all on project); and
- Secured lockup for evenings and weekends.

Safety reporting

Besides the general contractor's weekly inspections, OSHA or state safety agencies can perform their own inspections. These external inspectors can issue citations for safety violations and shut down operations considered to be life-threatening. Stiff fines may be levied for violations. Large construction firms are required to maintain detailed records for OSHA, but regardless of size, all contractors should do so. They are required to maintain a log of all jobsite injuries and illnesses and submit an annual report to OSHA at the beginning of each year. The jobsite injury log must contain the following information:

- Date of injury,
- Name of employee,
- Occupation of employee,
- Description of injury, and
- Lost time due to injury.

In addition to the log, contractors are required to submit a separate one-page description of each illness or injury. This supplemental record of occupational illness or injury includes a description of how the injury occurred, the place of the accident, the employee's activity at the time of injury, and the exact nature of the injury, including affected body parts.

Accident and near-miss investigation

Administration of safety processes is an important aspect of construction management. The creation of a safe working environment is a function of the physical conditions of the working environment and the behavior or working attitude of individuals working on the site. A safe working environment results in increased worker productivity and reduces the risk of injury. Accidents are costly, lead to disruption of the construction schedule, and require significant management time for investigation and reporting. There are many potential costs due to a construction accident, including workers' compensation and general liability insurance rate increases. Performing accident investigations and reporting their result is strongly recommended and is often a mandatory requirement by OSHA. However, accidents sometimes do not occur by chance. Near-miss events are indicators of accidents waiting to happen. As part of proactive safety management, contractors often implement *near-miss reporting systems*, which replicate (to a minor extent) some of the accident investigation procedures, to promote a safety culture among craftsmen through continual recognition and reporting of near-miss events.

All construction jobsite accidents should be investigated promptly by site management, usually by the superintendent and/or the project-specific safety manager. Moreover, an individual or independent inspection team may be needed for serious accidents, such as the collapse of construction in progress. These third-party safety professionals are used to conduct major accident investigations to avoid biased findings. Accident investigations have some common steps and intended outcomes. As part of the accident investigation, the initial step is to secure the scene to allow for the investigation to occur. This also means assessment of potential hazards and verification that it is safe to enter. The investigator will then collect the facts from witnesses and the project team to describe the incident and the sequence of events that led to it. This step is intended to achieve one of the major outcomes: define what happened. Next, the investigator will attempt to identify the root and contribution causes that will need to be addressed to avoid the same incident from reoccurring. This step is intended to identify why the accident happened. Finally, the investigator will evaluate and recommend procedures or policies to be adopted to minimize the potential for future occurrence. An accident investigation by the GC's site management is kept in the project file. Photographs should be included with the report to document conditions at the accident site. Some clients will require a copy of each accident report. Fatalities must be reported to OSHA within 8 hours, and in-patient hospitalization, amputation, or eye loss must be reported within 24 hours. Reports must be submitted to OSHA, as well as kept on file with the construction company. Figure 7.4 shows an example of an investigation report form.

Example

Assume your flagger was unfortunately struck by a passenger car, unrelated to the construction project. Some of the issues that the safety officer would be concerned with, possibly with assistance from the project engineer, would include:

- Has the employee been taken for medical attention and, if so, is he or she okay and can they be returned to work that day or in the near future?
- Who was the foreman in charge of the employee? Did he or she witness the accident and file a written report?
- Were the police called, and do we have a copy of the police report and report number?
- Do we have photocopies of the driver's license, insurance card, and registration?
- Have photographs been taken from all angles possible?
- Has our insurance carrier been notified?
- Do we have a replacement flagger in place, and has he or she been educated on this occurrence?

Use of technology

Technology affects safety in many different ways. First, new tools and PPEs are continually introduced by vendors, and it is up to the general contractor to learn and decide which tools would help improve their performance. Screening new tools goes hand in hand with reviewing the effectiveness of the tools being adopted. Recently, it was discovered that a major vendor of safety

Gateway Construction Company
2201 First Avenue
Spokane, WA 99205
509-642-2322

ACCIDENT INVESTIGATION FORM

Project Name: _____

Project Location: _____

  1. Name of injured person: _____

  2. Date and time of accident: _____

  3. Job title of injured person: _____

  4. Nature of injuries sustained by injured person: _____
_____

  5. Describe the accident and how it occurred: _____
_____
_____

  6. What was the cause of the accident? _____
_____
_____

  7. Were all company safety policies being followed? _____ If no, explain: _____
_____

  8. Was PPE required? _____ Was all required PPE being worn? _____ If no, explain: _____
_____

  9. Describe the job site conditions at the time of the accident: _____
_____

10. Had the injured person been properly trained to perform the assigned tasks? _____

11. Witnesses:_____

Prepared by _____ Date _____

Figure 7.4 Accident Investigation Report Form

gear had sourced latches for their fall protection kits from a manufacturer overseas. Where these latches had been rated, a series of near-miss events showed that their strength was questionable. As a result, several contractors were subject to a recall and forbidden to use any of these latches on their construction sites.

Other technologies that are changing jobsite safety management include the use of mobile apps for near-miss events and hazard monitor reporting. Reports and visual information from the site are often uploaded to central databases for corporate use, including tracking safety performance and documenting safety efforts. For instance, Autodesk BIM 360 Field app and web-based portal allows contractors to create checklists and capture visual information on site hazards that can be shared with the whole team or assigned for follow-up to specific team members.

Project engineering applications

Similarly to quality control, project safety implementation is ultimately the general contractor's superintendent's responsibility, often contractually and legally. Still, everyone working on a construction jobsite is responsible for safety, because we all have a moral responsibility to report a potential safety issue to site management for resolution. Field engineers and project engineers are often assigned to be the eyes and the ears for site management. In addition, project engineers can support the superintendent or project safety inspector in various tasks, including:

- Facilitating Monday morning safety meetings,
- Producing safety meeting notes,
- Coordinating receipt of project-specific safety plans from subcontractors,
- Collecting SDSs and making them available to all for review,
- Logging safety violations,
- Preparing pep talks on specific safety topics for the whole workforce or for craftsmen in supervised trades,
- Performing or facilitating safety observation walks,
- Performing safety inspections in support of the GC's on-site/off-site safety personnel,
- Reporting lack of safety compliance to the superintendent or safety officer, and/or
- Analyzing data for safety reporting.

Summary

Construction is a dangerous industry, and jobsites pose many hazards to construction craftsmen, as well as the general public around a site. To minimize the potential for construction injuries, OSHA has established national safety standards that are enforced through a jobsite inspection program. Because of the importance of safety management, most construction firms have developed safety programs and employ full-time safety professionals. The GC's project manager and superintendent are responsible for safety on the construction site. It is their contractual, moral, and ethical responsibility to enforce good safety practices.

Prevention of construction accidents is paramount to a good safety program. This is accomplished by identifying all hazards that are associated with each work activity and developing plans for eliminating, reducing, or responding to these hazards. This is known as job hazard analysis. Substance abuse can also result in accidents, so an effective substance abuse program is needed to remove individuals from the jobsite who are under the influence. Personal protective equipment should be provided, and successful enforcement minimizes the risk of jobsite injuries. Construction employees need to be informed of all hazardous chemicals that will be used on the project, their potential effects, and the emergency and first aid procedures. Project-specific safety plans are developed to identify hazards that may occur and specific measures necessary to reduce accidents.

Any accident that may unfortunately occur on a construction site should be thoroughly investigated to identify key learnings. The objective is to determine why the accident or incident occurred and identify procedures or policies to minimize the potential for future occurrence of similar accidents. A construction firm's safety record has a significant impact on its labor cost. Workers' compensation insurance rates are based on the history of claims submitted by the construction firm's craftsmen—safer contractors have lower rates!

Review questions

1. How does the EMR affect a contractor's labor costs?
2. What type of information would have to be reported to OSHA regarding a construction injury in which a carpenter broke her arm?
3. What are the three objectives of a construction accident investigation?
4. What is a project-specific accident prevention plan?
5. What are SDSs? Why are they on construction sites?
6. What are the four prime causes of injuries in construction?
7. What is OSHA's role in construction safety management?
8. Why is jobsite safety a significant construction management issue?
9. What is the purpose of a job hazard analysis?
10. What are three ways to reduce the risk created by a hazard on a construction site?
11. Why is substance abuse an important safety issue on a construction jobsite?
12. How are drug and alcohol abuse programs in the construction industry typically administered?
13. What are four types of PPE that are required on most construction projects?

Exercises

1. Prepare a job hazard analysis for the construction of the wood framing on the Rose Collective mixed-use apartment project.
2. During a weekly inspection, you find that a subcontractor has just brought a bucket of Hilti Smoke and Acoustic Spray CP 572 on site and is planning to use it. Once back at the trailer, you realize that the SDS is missing for this product. It is your concern to verify that first aid measures on site are adequate to deal with this material. Can you find the SDS for this material and verify what first aid measures are recommended?
3. Develop the section of the phased accident prevention plan for the I-90 project that covers mobilization and the early site preparation phase of the project.
4. What can you tell about a contractor who has an EMR of 1.7 and another with an EMR of 0.7?
5. Which construction trade is likely to have a higher workers' compensation base rate:
   a. Concrete laborer or administrative assistant?
   b. Drywall installer or pile driver (pile buck)?
   c. Electrician or window-blind installer?
   d. Operating engineer or field engineer?

References

BLS (2016, December 16). Census of Fatal Occupational Injuries Summary, 2015. Retrieved September 26, 2017, from www.bls.gov/news.release/cfoi.nr0.htm

OSHA (2017). OSHA Data & Statistics: Commonly Used Statistics. Retrieved September 26, 2017, from www.osha.gov/oshstats/commonstats.html

The National Institute for Occupational Safety and Health (NIOSH). (2016, January 18). Hierarchy of Controls. Retrieved September 26, 2017, from www.cdc.gov/niosh/topics/hierarchy/default.html

8 Production control and reporting

Introduction

Production control and reporting is an important element of construction project management. As stated in Chapter 2, cost and time are two critical dimensions of project success, with the others being quality and safety. Therefore, the jobsite project team is responsible for ensuring that the project is completed on schedule and within budget. To accomplish this challenging task, production controls—or cost and time controls—are established to monitor progress throughout the project. The project manager, superintendent, and project engineer must work together to ensure that all contractual requirements are fulfilled while a profit is earned on the project. Early completion of a project results in reduced jobsite general conditions costs, which can improve profits. However, acceleration of tasks may increase direct costs due to various issues, including inefficiencies, overtime, rework, and safety issues. This type of trade-off decision between the four pillars of project success is an advanced topic that will be discussed in Chapter 22. In this chapter, we will examine several techniques for cost and schedule control at the field level. Additional control processes with a focus on supporting project manager and home office are discussed in Chapter 12.

Cost control

Cost control begins with the assigning of cost codes to the elements of work identified during the work breakdown phase of developing the cost estimate. The development of a *work breakdown structure* will be discussed in Chapters 17 and 18 as part of estimate and schedule development. Cost codes allow the project manager, the project engineer, and the superintendent to monitor project costs and compare them to the estimated costs. The objective is not to rigidly keep the cost of each element of work under its estimated value but rather to monitor that the total cost of the completed project is forecasted to be under the estimated cost. Some uses of actual cost data include:

- Providing the project client with a cost report, which may be an open-book contract requirement;
- Evaluating the effectiveness of the jobsite project management team;
- Developing a database of historical cost data that can be used in estimating the cost of future projects;
- Monitoring project costs, identifying any problem areas, and selecting mitigation measures;
- Identifying additional costs incurred as a result of changes and process change order proposals (see Chapter 20); and/or
- Identifying costs for completing work that was the responsibility of a subcontractor and processing associated back charges.

After the bid or baseline estimate is established, the cost control process, as depicted in Figure 6.2, involves the following steps that were discussed in Chapter 6:

- *Cost codes* are assigned to each element of work;
- Once the bid or proposal is accepted by the client, the cost estimate is *corrected* based on buyout values. These are the actual values of the subcontracts and supplier bid packages. Project buyout will be discussed in Chapter 19;
- Actual costs are *tracked* for each work item by using the assigned cost codes;
- The construction process is *adjusted*, if necessary, to reduce cost overruns, which includes potential change orders and/or back charges; and
- Actual quantities, costs, and productivity rates are recorded, and an *as-built* estimate is prepared.

All jobsite costs must be monitored, but the items that primarily involve the greatest risk to the general contractor (GC) include:

- Direct labor,
- Equipment rental, and
- Project administration, also known as jobsite overhead or general conditions.

It is possible to lose money on material purchases, but with good estimating skills, it is not probable, and the risk is not as great as it is on direct labor. Costs of direct labor are based on estimated productivities, which can vary depending upon multiple factors. The same holds true with subcontractors. The subcontractors have quoted prices for specific scopes of work, and they therefore bear the risk associated with their own labor and equipment. However, a subcontractor's failure to perform may increase the cost of project administrative staff for the general contractor, including project engineers' time for overseeing and eventually assisting the struggling subcontractor.

Controlling costs is difficult if the project team does not start with a detailed estimate. For example, let's suppose there was a $1,014 cost overrun on the sidewalks of the I-90 case study. The aggregate cost analysis shown in the upper quadrant of Table 8.1 does not provide sufficient detail to identify the cause. The project cost control team should use a more detailed cost breakdown, as shown in the lower quadrant of Table 8.1, to determine the cause of the cost overrun.

Table 8.1 Aggregate and Detailed Cost Analyses

		Quantity	Unit	Estimated		Actual		Cost Variance
				Unit Price*	Cost*	Unit Price*	Cost*	
Aggregate Cost Analysis	Sidewalks	600	SY	$13	$7,800	$14.69	$8,814	−$1,014
Detailed Cost Analysis	**Sidewalks**	**600**	**SY**	**$13**	**$7,800**	**$14.69**	**$8,814**	**−$1,014**
	Edge formwork	1,350	LF	$1.00	$1,350	$ 0.80	$1,080	$270
	Buy concrete	66	CY	$80.00	$5,280	$78.00	$5,148	$132
	Place concrete	66	CY	$3.00	$ 198	$ 4.00	$ 264	−$66
	Finish walks	5,400	SF	$0.18	$ 972	$ 0.43	$2,322	−$1,350

* without markups

The detailed cost breakdown makes it easy to see that the problem is not with the formwork installation, concrete procurement, or concrete placement. Instead, the majority of the overrun is due to concrete finishing. Why did this overrun happen? Maybe the concrete mix was too wet or too dry. Maybe the sidewalks were rained on before they could be finished. Maybe personnel changes were necessary. Maybe the estimator assumed a broom finish, but a more expensive hard-trowel finish was required. There could be a variety of reasons. Use of a detailed cost breakdown allows the jobsite project team to identify specific issues, so they can focus on evaluating them.

Cost codes

To be able to control costs, they must be tracked accurately and compared against the buyout-corrected estimate. The first step is to record the actual costs incurred and input the information into a cost control database. Cost codes are used to allow comparison of actual cost data with the estimated values. There are several types of cost codes used in the industry. The best system to use on most projects is the coding system selected for the project files. Many construction firms have adopted the Construction Specification Institute (CSI) MasterFormat system, but some project owners may require a different system. An example of this type of coding system is:

$$\overline{\text{(project number)}} \ \overline{\text{(CSI work package)}} \ \overline{\text{(element of cost)}}$$

The project number is assigned by the construction firm, the CSI work package code is from the MasterFormat system, and the element of cost is the type of cost. An example of element of cost coding is:

1. Direct Labor
2. Equipment Rental
3. Direct Material
4. Subcontract

With this system, the cost code for steel beam placement direct labor on the I-90 bridge overpass (project # 1732) deck would be:

1732.05-32-00.1

The cost code for redi-mix concrete purchase for the bridge deck would therefore be:

1732.03-32-00.3

Depending upon the size of the construction firm, the type of work, and the type of client and contract agreement, the general contractor may perform job cost accounting in either the home office or in the field. Generally, the smaller the firm and the smaller the contract value, the more likely that all accounting functions will be performed in the home office. On larger projects, the project team may have a jobsite accountant, even if automation and information technology is reducing the need for decentralized accounting. However, the type of contract and how it addresses reimbursable costs may also have some effect on where the construction firm performs the accounting function. Let's say, for example, the project is a $200 million aerospace manufacturing facility that has a negotiated guaranteed maximum price (GMP) contract that allows for all on-site accounting to be reimbursed. Whereas it may be more cost-effective to perform accounting out of the home office with the assistance of the accounting department, according to the terms of the contract, the client will not pay for activities conducted off the project site; therefore, an accountant is required in the jobsite trailer.

Regardless of where the cost data are collected and where the checks are prepared, most of the accounting functions on a project are the same. The process begins with a buyout-corrected estimate. Actual costs are then incurred, either in the form of direct labor, material purchase, or subcontract invoice. Cost codes (those matching the file system and the estimate) are recorded on the time sheets and invoices. Often recording of cost codes is initiated and performed by the project engineer. The coded time sheets and invoices are then passed to the superintendent and project manager for approval. Sometimes, the officer-in-charge (OIC) or the project owner (on cost plus projects) may also want to initially approve each invoice. After the time sheets and invoices are coded and approved, the cost data are input into the cost control system.

One important aspect of this phase of cost recording is the accurate coding of actual costs. If costs are accidentally or intentionally coded incorrectly, the project team will not really know how they are doing on that specific item of work. Some superintendents may have their project engineers intentionally code costs against items where there is money remaining, rather than against the correct work activity, thereby hiding overruns. With this approach, the jobsite staff will not be able to monitor and collect accurate cost data if coding errors occur. All costs should be coded correctly to provide the project team with an accurate accounting of expenditures.

Work packages

Control or management of direct craft labor and equipment rental cost is the responsibility of the superintendent, often with assistance from a field engineer or project engineer (PE). The key to getting the foremen and superintendents involved in cost control is to get their personal commitment to the process. One successful way for the general contractor to do this is to have the superintendent actively involved in developing the original estimate. If the superintendent said it will take six ironworkers working three days to plumb the structural steel, he or she will endeavor to see to it that the task is completed within that time.

Some construction executives believe that superintendents should not be told the true budgeted value of each work package. This is a poor practice, because the foremen and superintendents are key members of the project team and have critical roles in making the project a success. They should be provided with the actual budgeted cost for installation, both in materials and man-hours, as well as the scheduled time for installation, for each work package.

A good technique for monitoring project cost is to develop a direct-work project labor curve similar to the one shown in Figure 8.1 for the Rose Collective case study. For this example, all of the direct work has been combined into one curve; it is not separated by craft or work package, as will be discussed below. It is important to have the superintendent or foreman record the actual hours incurred weekly and chart them against the estimate; the PE can help with this. If the actual labor used is under the curve, the foremen and their crews are either beating the estimate or behind schedule. The opposite is true if the hours are above the curve. This simple method of recording the man-hours provides immediate and positive feedback to the project team and is, therefore, usually preferred to monitoring dollar expenditures. The advantage of estimating with unit man-hours (UMH) over unit prices for labor is that foremen and superintendents think in terms of crew size and duration; they do not think in terms of $13 per square foot of sidewalks.

Many in the construction industry, especially estimating consultants to the project owner or designer, think in terms of organization, according to the strict CSI MasterFormat. As a result,

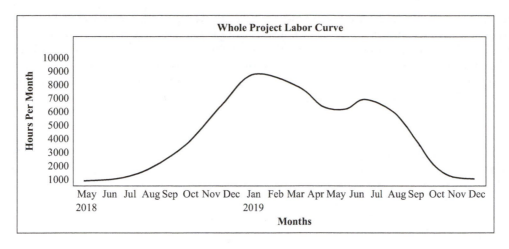

Figure 8.1 Project Labor Curve

since CSI division 03 is for concrete, all concrete work is lumped together and not formatted in the estimate according to systems or assemblies. Therefore, all concrete foundations, slabs on grade (SOG), columns, elevated decks, and walls are grouped together. This may even result in grouping off-site pre-cast concrete with on-site cast-in-place or tilt-up concrete. Conversely, any work that may be specified in another CSI division, such as structural excavation and backfill or footing drains or structural steel embeds, will not be grouped with the concrete work, even though these activities are performed together.

However, contractors do not organize their estimate by relying blindly on the CSI MasterFormat. Preparation of the work breakdown structure (WBS), as described in Chapters 17 and 18, is also more efficiently done by work packages or assemblies than a pure CSI approach. Work packages are a method of breaking down the estimate into distinct packages or assemblies and systems that match measurable work activities. For example, footings—including forming, reinforcing steel, and concrete placement—could be a work package. The work is planned according to the number of hours in the estimate and monitored for feedback. When the footings are complete, the superintendent and management team will have immediate cost control feedback. Work packages apply best to those who estimate and track costs by system, rather than a pure CSI approach. When a system is complete—such as footings, slabs on grade, or steel columns—the project team immediately knows how they are doing with respect to the overall estimate.

One of the most evident advantages of planning construction by assembly or work packages, in lieu of CSI, is the ability to monitor progress in both time and cost. When an assembly such as punch window installation, which is part of CSI division 08, is complete, the contractor will know exactly what the windows cost and how long it took for their installation. He or she will then have immediate feedback as to how the original budget and schedule look. If all of CSI Division 08 was grouped together, including windows, storefront, interior and exterior doors, frames, door hardware, and access doors, the contractor would not necessarily know how the budget was looking until all of this work was complete.

Which items should the project team track? The *80–20 rule* applies here and elsewhere in construction management. Eighty percent of the cost and the risk lie with 20 percent of the activities. Therefore, those activities deserve the most attention in cost control. The jobsite project team should evaluate and mitigate the largest risks. The estimate should be reviewed to identify those items or systems that have the most labor hours. Work packages should be prepared for those items that the cost control team believe are worthy of tracking and monitoring. Each work package should be developed by the foreman who is responsible for accomplishing the work, maybe with assistance from the project engineer. Some examples of simple straightforward assemblies that warrant work packages on our apartment case study project include:

- Spot footings,
- Cast-in-place (CIP) concrete elevator pit,
- Garage SOG,
- Wood backing,
- Roof accessories,
- Doors/frames/hardware,
- Toilet accessories, and
- Kitchen appliances.

Contractors think of work in terms of assemblies or systems, which are based on the selected workflow and not strictly on the CSI Masterformat classification. The concept of work packages involves grouping all items of work into one assembly, regardless of CSI, supplier, or craft installation. In this way, the contractor knows what that assembly costs and where it fits into the project schedule. Decisions on the means and methods of construction belong to the contractor, and the superintendent leads these decisions. The superintendent also efficiently assembles all the materials, labor, and equipment necessary to construct the work.

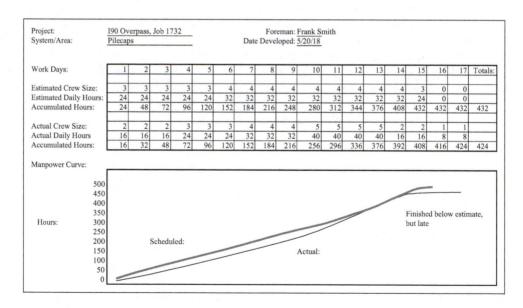

Figure 8.2 Foreman's Work Package

Example: Work package for concrete pile caps

The superintendent is meeting with his or her foreman to plan the work and hand over the work package for assembly. To envision an effective operation, the superintendent will consider:

- The auger-cast piling subcontractor's schedule and contact information (i.e., when the auger-cast piling subcontractor is expected to be on site, and who the subcontractor's contact person is to expedite or get insurance papers from);
- The foundation plan and details;
- Approved installation submittal drawings and schedule for reinforcement steel delivery;
- Any requests for information (RFIs) processed for this work;
- Contact information for the backhoe subcontractor who will clean up the auger-cast spoils and excavate and backfill around the pile caps and grade beams. If off-site material is needed for select backfill, then supplier contact information for the approved material is also needed;
- If plumbing or electrical underground work was necessary prior to the concrete work or sleeves were provided through the concrete, then coordination with those subcontractors is necessary;
- Delivery schedule and laydown for formwork material, whether the forms are to be job-assembled or rented from a supplier;
- Availability of tools and equipment, such as forklift, welder, compressor, etc.;
- Anchor bolts and templates for the structural steel columns, which will sit atop the pile caps;
- Assignment of a surveyor to control grades and lines;
- Schedule allotment from the detailed contract schedule, such as three weeks;
- Estimated installation hours from the estimate, crew size, and personnel selections, such as three to four men working for three weeks for a total 432 man-hours;
- Approved submittal for concrete redi-mix design, along with an estimate of 140 cubic yards of concrete required;
- Contact for the third-party inspectors for rebar, concrete slump, and backfill compaction; and
- Most commercial and civil concrete is pumped; therefore, the availability of a pump subcontractor, along with pump size, and the rental rates and contact information.

There are likely additional details and information that the foreman will need, but this is just an example of all the elements that might go into one simple work package. Work packages for systems—such as elevated concrete decks, structural steel, wood framing, siding, or roofing—would be even more involved, but this is all necessary to prepare a thorough work package to assure a successful construction operation. Figure 8.2 shows an example work package cost control worksheet. The detailed estimate for this system is shown in Chapter 17.

Schedule control

Schedule control, or management, involves monitoring the progress of each activity in the construction schedule and determining the impact of any delayed activities on the overall completion of the project. Schedule control is just as important as cost control, because the project team must ensure substantial completion is achieved prior to the required contractual completion date. If this does not occur, liquidated damages may be owed to the client to compensate for the project not being completed on time and extended jobsite general conditions will quickly erode any planned profit. Proper schedule management is very similar to cost control. First, the field team needs to be given all of the detail to properly do their job controlling the schedule. Then, superintendents utilize information from the general schedule to develop *short-interval schedules*, also known as *look-ahead schedules*, that provide details on the workflow and are used to continually monitor progress, and communicate with direct labor and subcontractors. Look-ahead schedules are frequently developed and updated, and their interval may be two, three, or four weeks depending on the project complexity and intensity. Meanwhile, project engineers in the field and foremen observe work advancement and feed the superintendent with up-to-date snapshots on field activities in the form of completion and advancement logs. An example of a graphical log for a high-rise building project is provided in Figure 8.3.

These look ahead schedules and completion logs have a different purpose from the schedule updates and development that will be discussed in Chapters 16 and 18, respectively. Master schedules and their updates look at the project as a whole with the target of completion. On the other hand, short interval schedules and their updates look at providing quick feedback to highlight issues as they occur and make corrective actions to avoid these issues. This may substantially impact the master schedule, as well as the performance of the project in terms of quality, safety, and cost. For instance, the completion log in Figure 8.3 shows a few issues that may need to be addressed immediately. First, the drywall subcontractor has either left behind or encountered conflicts with its work in one apartment in Level 27. Similarly, the casework subcontractor has not completed installation of the kitchen and vanity cabinets in a few apartments on Level 29. Lastly, the flooring subcontractor seems to be holding everyone up on Level 28. These pieces of information would be useful agenda items for the foreman's meeting and would eventually justify individual meetings with site supervisors from these subcontractors. The superintendent will then report construction progress at the weekly project coordination meetings when the jobsite project manager will review the progress with the project team. The project manager will lead a discussion on the causes of any delay or issue. Moreover, the project manager should select appropriate mitigation measures with the project engineer and superintendent. Such measures may include expediting material delivery, increasing the size of the workforce, working extended hours, facilitating issue resolution by increasing close-up coordination and supervision of specific subcontractors, or making changes to the GC's own processes that have failed or are inefficient. Updated short-interval schedules will be developed to intensively manage all accelerated work. These are all activities that the PE can assist the superintendent with.

Use of technology

The activities described in this chapter may benefit from various technological advances. Some construction firms rely on integrated technologies for performing all steps. For instance, some contractors have their own *enterprise resource planning* (ERP) platform, which is used to manage accounting, billing, safety reporting, and other processes. The costing application is pre-populated with cost codes, which follow corporate guidelines. The project management team will review these pre-populated codes and adapt them to the project. This approach is that followed by other industries where automation is taking over many tasks. However, many contractors still use technology to support only some steps or each step independently and rely on manual entry for transferring information from one step to the other. Small contractors may still perform all of these tasks manually, but technology is reducing the need for manual coding. Still, project staff should be able to review codes and understand them, so that specific items in subcontractor invoices can be linked to budget line items. For instance, if an invoice includes plywood, should it be charged to formwork or temporary structures?

Figure 8.3 Sample Finishes Completion Log

Floor Level / Apartment Unit

Task/Party Responsible	27 (A–M)	28 (A–M)	29 (A–M)
Route out wall safe/GC			
Install drywall/Sub 1			
Install kitchen cabinets/Sub 2			
Drill appliance holes/Sub 2			
Install vanity cabinets/Sub 2			
Install tub tile surrounds/Sub 3			
Grout and/or caulk tub tiles/Sub 3			
Install flooring/Sub 3			
Install kitchen countertop/Sub 4			
Install vanity countertop(s)/Sub 4			
Grout and/or caulk countertop/Sub 4			
Install bases and door frames/Sub 2			
Hang interior doors/Sub 2			
Install roller shade backing and valances/Sub 2			
Install backsplash tile/Sub 3			
Grout and/or caulk backsplash tile/Sub 3			
Paint bases and door frames/Sub 5			
Caulk window wall to drywall/Sub 5			
Install entry doors/Sub 2			
Install door hardware/Sub 2			
Install bath accessories/Sub 2			
Install closet shelving/Sub 2			
Install plumbing trim/Sub 6			
Install electrical trim/Sub 7			
Install mechanical trim/Sub 6			
Install mirrors and shower doors/Sub 8			
Install roller shades/Sub 9			
Perform final paint/Sub 5			
Install appliances/Sub 10			
Construction cleanup/GC			
Pre-Punch List/All			
Owner Punch List/All			
Unit Locked Out/GC			

Legend:

Not available for inspection	0
Not started	1
Work in progress	2
Completed	3

As for all controlling and monitoring activities in the field, the workflow is based on a three-step approach: (a) performing continual field observations and measurements; (b) documenting and archiving information into a repository, such as a physical file cabinet or an electronic database; and (c) extracting information from the repository and analyzing it to provide up-to-date reporting on how the project field activities are faring against the budget and the schedule plans.

Field observations and measurements

In the field, a broad assortment of technology tools is available to project engineers to input what they are observing. Depending on what the project engineer wants to record, different tools may be used. However, the type of information to be recorded may vary in scope and detail. Whereas visual observations were converted into a narrative in the past, they are now frequently captured through standard or 360-degree digital photography. If ground-level visual observations are not possible, they can be enhanced with drones that take photos from the sky. Drone photography is especially useful to visually represent work advancement in heavy civil projects. If visual information is not sufficient, laser scanners are used to provide dimensional information.

Repository for field observations and measurements

Most information collected on field production is frequently recorded electronically into a database. However, some of the input from the field may have been recorded on paper. In this case, the project office will probably have a file cabinet where the paper documentation is recorded and archived. Whereas documents in this cabinet once served as the permanent project repository, the cabinet now mostly serves as a temporary repository until a project engineer is charged with reviewing the information and inputting it into an electronic database. The paper version is then scanned and uploaded into the database. Electronic repositories used to be databases stored locally on a computer or server in the project office. Larger companies now also have corporate databases that are often web-based and can be accessed on computers, tablets, or mobile phones through an internet browser.

Example No. 1: Produce weekly completion log on a high-rise project

A project engineer has been tasked to develop a weekly update to the finish completion log, such as that shown in Figure 8.3. To perform this task, the project engineer will need to walk all levels and apartment units where finish tasks have been started, according to the master schedule. The goal of the completion log is to report to the superintendent and project manager on how activities are proceeding, if delays have occurred, and if issues have arisen in any place in the building. Information on any task for any apartment unit by floor level could be inputted into a tablet through a checklist or an electronic form, which would be electronically processed to produce a completion log similar to Figure 8.3. If a tablet is not available, the project engineer could still perform the task by recording his or her observations on paper on a clipboard. Obviously, this approach would add another step to the project engineer's workflow because he or she would need to enter the data into a computer system to produce the log. Moreover, data entry errors may occur when transferring the information from paper to the computer system for analysis.

Information extraction for report generation

All electronic databases allow users to query the repository to create specific reports. Initially, this feature was only available toward the end of the process. However, the recent advent of mobile and cloud computing has resulted in the appearance of multiple mobile applications that integrate the processing steps. Therefore, project engineers can input information in the field directly into the repository, while they can also run queries from the field for quickly evaluating what they are observing.

Example No. 2: Produce daily reporting on floor installation

A project engineer is tasked with completing a daily report on floor installation for a new school building. Floor plans are available in portable document format (PDF) format. The project engineer is provided with a tablet with Bluebeam or some other application that allows area take-offs. Using the tablet, the project engineer observes the areas where floor installation has been completed, highlights them on the floor plans, and uses the take-off feature to compute quantities in place. These quantities are inputted on an electronic form that is linked with a Microsoft PowerBI tool, which produces a comparison between the actual and estimated production totals. If the floor installation was self-performed by the project engineer's employer, the project engineer could also obtain man-hours from the flooring foreman. This additional information would also allow a comparison between the actual and estimated costs of the floor task.

Project engineering applications

Project engineers have an opportunity to be actively involved in field production control and reporting on both heavy civil and building projects. As stated elsewhere, they can assist with:

- Estimating, including quantity take-off (QTO) and pricing direct work,
- Buyout of subcontractor and supplier bid packages,
- Drafting subcontracts and purchase orders,
- Cost coding supplier and subcontractor invoices,
- Charting whole-project labor curves, and/or
- Establishment and maintenance of foremen work packages.

The project engineer does not do any of this work in a vacuum. All of it will be done with the mentoring and oversight of more experienced project team members, such as the chief estimator and scheduler, the project manager, and the superintendent, along with in-field forecasting and reporting from the craft foremen—the people ultimately responsible for execution of the work. The project engineer can play a valuable supporting role in several of these activities. In addition, the PE plays a key part in improving the company's estimating database through input of accurate as-built estimates. The size of the project team also plays a role in what a PE may be asked to do. On big projects, a senior PE would review subcontractor invoices and submit them to the PM for final review. On a small- to medium-sized job, the PM may perform these tasks.

Summary

Production controls, including cost and schedule control, are essential project manager, project engineer, and superintendent functions. The project team must ensure that the project is completed within budget and on schedule. Cost codes are used to track project costs and compare them with the budgeted (or estimated) value for each element of work. Work package analysis provides the jobsite project team with a method for tracking the direct labor cost, which represents the greatest risk on the project. Schedule control involves monitoring the progress of each scheduled activity and selecting appropriate mitigation measures to overcome the effects of any schedule delays. However, the control and monitoring of field production relies on: (a) performing continual field observations and measurements; (b) documenting and archiving information into a repository, such as a physical file cabinet or an electronic database; and (c) extracting information from the repository and analyzing it to provide up-to-date reporting on how the project field activities are faring against the budget and the schedule plans.

Review questions

1. Assuming standard CSI specification sections, what would the cost codes be on the Rose Collective mixed-use apartment project (# 414) for:
    a. Second floor apartment wood framing labor,
    b. Gypsum wall board (GWB) subcontractor, and
    c. Kitchen appliance purchase?
2. What are four uses of actual project cost data?
3. What are the three steps necessary to successfully control and monitor field production?
4. Why is the original cost estimate corrected based on buyout data?
5. What three types of project costs present the greatest risk to the jobsite team?
6. What are project labor curves used for?
7. What is a foreman's work package used for? What are the advantages of planning the construction operations with work packages, rather than a pure CSI approach?
8. What are two other terms used for grouping of construction operations into work packages?
9. Other than the examples listed earlier, what are three assemblies or systems that might warrant the development of a foreman's work package on a typical project?

Exercises

1. What might be missing from the pile cap work package assembly discussed above?
2. Prepare a list of items and considerations for either: (a) a structural steel work package assembly, or (b) a wood framing work package.
3. Some contractors do not share their true estimates and schedules with their foremen. Prepare an argument for (a) why it is a good idea to share all this information and, conversely, (b) why it is a bad idea to share this information with the foremen.
4. Planning construction by work packages is not limited to general contractors. By using our Rose apartment case study as an example, break down the work for either the plumbing subcontractor, the heating, ventilation, and air conditioning (HVAC) subcontractor, or the electrical subcontractor into at least five distinct work packages or assemblies or systems.
5. List five reasons why a heavy-civil GC would be more inclined to perform work directly with their own crews versus hiring subcontractors. Simply saying better cost, schedule, quality, and safety control are not enough.

6. List five reasons why a commercial GC would be more inclined to perform work with subcontractors versus hiring their own crews. Simply saying better cost, schedule, quality, and safety control are not enough.
7. Prepare a list of the top ten systems or assemblies from our Rose apartment case study project that deserve the most cost control attention and fall within the 80–20 rule.
8. Provide an explanation of how and why the pile cap work package in Figure 8.2 followed the course it did and turned out in that fashion.
9. By utilizing another case study project, prepare a work package for direct labor for a system of work, such as concrete foundations, cast-in-place concrete walls, tilt-up concrete walls, or slabs on grade. Attach the original portion of the estimate that refers to this work package. Assume the following percentages complete and spent:

	Portion of Work	Portion of Estimate
	Complete	Spent
Excavation:	100%	110%
Form work:	95%	85%
Reinforcement steel or mesh:	90%	80%
Concrete placement:	40%	50%
Slab finish (if appropriate):	50%	65%
Tilt-up panel erection (if appropriate):	20%	15%

9   Quality control and reporting

Introduction

The concept of quality will differ depending on the industry. Furthermore, different project owners will interpret quality differently and often adapt their quality expectations to the project. In Chapter 5, an example was introduced on how a bank would probably have different expectations when building a suburban branch office than when building its new headquarters. For the headquarters, the bank may want to use the building as a way to convey the financial soundness of the company. For the suburban branch, the motivation could be based on value by optimizing the "bang for the buck" while satisfying the users' needs. Independent from the owner's viewpoint on quality, a construction firm will use the contractual documents to identify the owner's quality expectations. Therefore, it is paramount for the project owner to convey their expectations through these documents. During the project delivery, however, the contractor can identify gaps in the contractual documents that could result in quality shortfalls. Chapter 10 will discuss how gaps can be identified and addressed.

A contractor's approach for managing quality is an important element of construction project management. As stated in Chapter 2, quality is one of the four critical dimensions of project success, with the others being cost, time, and safety. Shortfalls in quality performance may have serious implications and affect the time, cost, and profit on a project. If quality issues need to be addressed, a contractor has to allocate project resources to perform rework, which may delay project completion and will increase costs. Quality shortfalls also offer long-term implications by affecting the overall reputation of the construction firm. The construction firm's greatest marketing assets are satisfied clients, and delivery of quality projects is critical to achieving customer satisfaction.

All individuals in the contractor's project team must work together to ensure that all materials used and work performed on a project conform to the requirements of the contract plans and specifications. Non-conforming materials and installations must be replaced at the contractor's cost, both in terms of time and money. This means that the contractor must bear the financial cost of tearing out and replacing the non-conforming work and that additional contract time is not granted for the impact the rework has on the construction schedule. Quality management is essential to ensure all contract requirements are achieved while minimizing rework. Similarly to safety control, all members of the project team, not just the superintendent, need to proactively manage the quality control process of the construction project.

Project-specific quality control planning

The general contractor's (GC's) jobsite project team must ensure quality materials are procured and received, quality craftsmen and subcontractors are selected, and all workmanship meets or exceeds contract requirements. This proactive approach to selecting effective procedures and processes to deliver quality projects can be pursued through different management systems, such as *total quality management* (TQM), and *lean construction*. All these management systems rely on careful planning that outlines an effective inspection process to ensure quality results. Project quality plans include both quality assurance (QA) and quality control (QC) processes. QA processes outline the methodology to be followed to assure quality results and tend to be similar across projects of the same type. QA processes are based on predicting what can go wrong in order to outline QC processes that would allow the contractor to find quality issues, if any are present. Therefore, quality plans identify how variations from standards are measured through quality control, how to ensure quality control is occurring according to the QA processes, and when actions should be taken to correct all variations or deficiencies. Effective quality planning starts with a detailed study of the contract documents to determine the project's unique quality and testing requirements.

Quality plans are developed during the pre-construction phase or project start-up to document the QA/QC procedures to be used on the project, which is a similar approach to developing project-specific safety plans, as was discussed in Chapter 7. Whereas QA processes are usually identified at the company level, adapted for the specific project, and implemented through external audits, QC procedures included in the project-specific quality plan are developed and implemented by the jobsite team. We will refer to the project-level QC processes in the rest of this chapter.

Most project owners require submission of quality plans, but the submission timing varies greatly, depending on the procurement method and the type of owner. Some private owners require quality plans to be submitted with the cost and schedule proposals, along with project-specific safety plans, during contract negotiations. Submission of quality plans may also be requested as part of the proposal under best value and quality-based selection. When quality is not a criterion for selection, such as under competitive bidding for public works, the project owner may still require QA/QC plans to be submitted by the successful GC at the pre-construction meeting.

QC processes are a prime example of "active" versus "passive" quality management. Active quality management is defined as taking measures to prevent mistakes and assure that the project goes well the first time around, rather than repairing non-conforming work after the fact. A project-specific quality plan is developed that identifies QC systems for each major work package. To manage quality on a project, a contractor will need to implement several tasks as part of their QC plan, including:

- Studying the prime contract agreement for contract quality expectations,
- Reviewing and adapting QA processes,
- Developing a site-specific QC plan that includes clear expectations and roles for all team members, including the project owner and designers,
- Submitting the site-specific quality plan to the project owner and design team for review and approval,
- Employing best value quality subcontractors,
- Requiring subcontractors to also submit project-specific QC plans,
- Utilizing in-house QC resources/specialists supplemented on an as-needed basis by third-party inspectors,
- Reviewing the design for gaps and promptly submitting requests for information (RFIs) to the project owner and design team for review and approval,
- Inspecting materials before unloading,
- Developing pre-installation mock-ups,
- Conducting in-process pre-inspections,
- Documenting non-conforming work and seeking remedy of any through contractor's punch lists,
- Performing formal punch list inspection and resolution, and
- Preparing for post-project warranty management (see Chapter 14).

Implementation of the quality control plan

Implementation of the QC plan requires that only specified materials are procured and qualified subcontractors are employed. Shop drawings and certificates of compliance with reference standards submitted by the suppliers and subcontractors must be carefully reviewed to ensure that all materials used on the project meet contractual requirements. During the review, the contractor project team may identify gaps in the contractual documents; these would need to be discussed at the weekly meeting with the owner and design team and may need to be addressed and documented through a formal RFI. Individuals assigned the duty of performing quality inspections should also participate in the review and submittal process, so they can become familiar with procured materials. Similarly, individuals assigned the duty of receiving materials should compare delivered materials against submittals and contractual requirements.

In addition to oversight of the general contractor's own direct crafts, the jobsite team must manage the quality of the subcontractors. Because subcontractors often comprise more than 80% of the work on a building project, ultimate quality success depends upon these valuable team members. Proactive project-specific quality plans from the GC and each of the subcontractors are essential toward achieving that goal, in addition to extensive testing and inspections of all direct and subcontracted work.

Subcontractors

The general contractor's reputation for quality work is greatly affected by the quality of each subcontractor's work. Quality subcontractors are not necessarily the least-cost subcontractors, but in the long-term, they will deliver a superior project. The term "best value subcontractors" refers to those who will deliver an optimized service in terms of cost, time, safety, and quality. This can be visualized through this formula:

Best Value Subcontractors = f (Cost, Time, Safety, Quality)

Before scheduling each subcontractor to start work, the project superintendent must ensure that their area of the site is ready for them. Quality requirements should be emphasized at subcontractor pre-construction meetings. Mock-ups, which are standalone samples of completed work, should be planned to establish the workmanship standards for critical features. Mock-ups are likely required to be constructed prior to actual work to establish QC standards. Subcontractors may be required to prepare mock-ups for exterior or interior finishes to establish the required level of workmanship. Figure 9.1 is an example of a mock-up of mechanical equipment. By developing a mock-up, a subcontractor will fully understand the quality requirements before being allowed to start work on the project. An example of a mock-up for the Rose mixed-use case study project would be a separate, standalone, apartment unit. This could be built out in adjacent warehouse space, including furnishings. Mock-ups for apartments, if properly accomplished, can later be used for pre-leasing showrooms. Subcontractor mock-ups are examined and often punch listed, i.e., reviewed against contract documents for compliance as a part of the preparatory inspection for each phase of the work.

Tests and inspections

Proactive quality control starts with selection of best-value subcontractors and a thorough submittal process. Approved materials and equipment are then procured by using the methods described in Chapter 19. All materials and equipment should be inspected as they are being

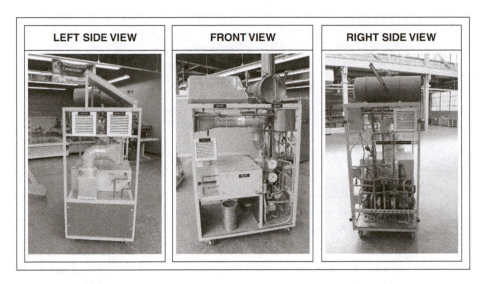

Figure 9.1 Example of a Mock-Up

delivered to the project site. They should be examined and compared to their purchase orders and submittal documentation; this duty is often assigned to a project engineer (PE). If the materials or equipment are not as specified, they should not be unloaded from the truck and should be rejected. Unfortunately, issues with materials or equipment may be detected too late to reject delivery. Under these circumstances, the project engineer should at least describe any deviation from the process on the paperwork and request their replacement. Materials on site and waiting for installation need to be re-inspected weekly to ensure they are safe and protected against the weather and damage.

Full-time in-house GC-provided QC inspectors may be specified for large or complex projects, such as hospitals. Some contractors attempt to substitute a foreman or assistant superintendent who already has other responsibilities. If a full-time inspector is specified in the contractual documents, the general contractor must provide one and include the cost in the general conditions part of the estimate. Qualified inspectors should be chosen to ensure that all self-performed and subcontracted work meets contract specifications. On less complex or smaller projects, the quality control and safety inspection functions may be performed by the GC's superintendent, foreman, or project engineers or by home-office staff specialists during weekly visits. Management of the jobsite quality process includes establishing standards in specifications and contracts, inspections, and necessary rework. Checklists are developed and included in the QC plan as tools for inspectors to check for critical items. An example QC checklist for the vinyl plank flooring for the Rose apartment case study project is shown in Figure 9.2.

Inspectors conduct initial inspections to ensure workmanship meets contractually established quality control standards. Daily follow-up inspections are conducted to ensure continuing compliance with contract requirements. Building inspectors and design team members should be scheduled to visit the site throughout the course of the project to ensure that all work is inspected prior to being covered by other work items. These inspections should ensure that all preliminary work has been completed and accepted before the next subcontractor begins its work. Subcontractors should not build over work that has been improperly installed by a previous subcontractor. Doing so constitutes acceptance of an unacceptable substrate.

Reliable Construction Company
1401 Waterfront Avenue
Portland, OR 97210

QUALITY CONTROL CHECKLIST

Spec: _____09680_____   System: Vinyl Plank Flooring, First Floor Only

Date: 9/27/2019   Conforms?

Project: Rose MXD Collective, Job 414

		Yes	No
1	Verify gypcrete moisture content is acceptable	Y	
2	Clean subsurface before application	Y	
3	No other subcontractors in area at this time	Y	
4	No exposed glue shown		Minimal
5	No turned-up edges or corners		No
6	No scuffs on tiles	OK	
7	No gaps between tiles	OK	
8	Finish close to wall to allow base to be covered	w/I 1/8"	
9	Direction consistent (N-S) or (E-W)	N-S	
10	Transition strips	Y	Replace 1
11	Surplus material left for owner's use		Open
12			

Notes: Subcontractor will correct deficiencies within one week

Surplus materials will be turned over after the fourth floor is completed

Inspector(s): David Arnold/QC, Joseph Wilson/Foreman, Jennifer Forsythe/PE

Figure 9.2 Quality Control Checklist

The owner–GC prime contract for most projects should address uncovering work for inspection. Work covered contrary to the architect's or engineer's request or in conflict with contract requirements is to be uncovered at the contractor's expense if deemed unacceptable. The design team may request that additional work also be uncovered for inspection. For instance, building codes require that plumbing pipes and electrical conduits receive a coverage inspection before gypsum wall board is installed; similarly, site utilities are inspected before they are covered with soil. The contract often includes language that allows the owner or designer to ask for uncovering work components, if they are not certain of what has been installed. If installed materials are found not to conform to contract requirements, the re-inspection and correction costs are then the contractor's responsibility. If the work is instead found to be according to the contractual requirements, then the owner pays for the inspection.

Contract specifications should identify work items that require testing and inspection. Whereas some owners could perform inspections with their own forces, it is common for third-party professional testing and inspection agencies to be engaged to perform the testing. The project owner usually employs these firms directly, outside of the GC's control, to guarantee their independence. Oftentimes, this arrangement is a permitting requirement from one of the authorities having jurisdiction (AHJs), such as the municipality where the project is located. For a building project, the owner–GC prime contract agreement often requires independent third-party testing and inspection of:

- Underground utility backfill, soil compaction, and shoring systems,
- Reinforcing steel and post-tension cables,
- Cast-in-place and pre-cast concrete,
- Structural steel fabrication and field welding,
- Fireproofing, roofing, and waterproofing,
- Glazing, windows, and curtain-walls, and
- Elevators.

In addition to inspecting materials and installed assemblies, operational systems testing, balancing, and commissioning are scheduled to ensure that all mechanical, electrical, and plumbing or piping (MEP) systems meet contractual performance criteria. MEP testing is discussed in Chapter 14.

If a proactive quality management system has been used on the project, the GC will perform pre-punch inspections on its self-performed and subcontracted work. Project engineers are often delegated this task. These inspections will result in the identification of issues and the communication of them to subcontractors with a timeline for resolution. For instance, once a subcontractor has completed its scope for the apartments in the mixed-use case study project, a project engineer would walk the apartment and use colored tape to highlight imperfections in painting, casework, and appliances and fixture installation. If the inspection is intended to highlight issues with multiple subcontractors at once, tape of different colors will make it easier for each subcontractor to identify their issues before the owner's punch list inspection is conducted. Examples of issues that can be raised during these inspections are shown in Figure 9.3. Some of the examples refer to an individual issue and are self-explanatory:

- Example A: It appears that the electrician left out one of the wires during the installation of the pendant lighting fixture. The electrical subcontractor should address this issue.
- Example C: The dryer vent duct was installed without using the hole in the shelf. The appliance subcontractor should address this issue.
- Example D: A plumbing trim is missing in the washer/dryer closet. The plumbing subcontractor should address this issue.
- Example F: The power cord for the electric oven was passed through the wrong hole in the cabinet. The appliance subcontractor should address this issue.

Other examples are more complex and may be due to one issue causing the others:

- Example B: An imperfection in the drywall was patched but not sanded. The drywall subcontractor should address this issue. Also, final painting was not completed. This issue seems to be caused by the drywall issue. The painting subcontractor should address this issue once the drywall contractor has addressed the drywall issue.
- Example E: There are several scratches on the flooring and marks on the countertop. These issues seem to be a result of *trade damage* and should be addressed by the countertop and flooring subcontractors. However, what is the oven appliance doing in there? After a close

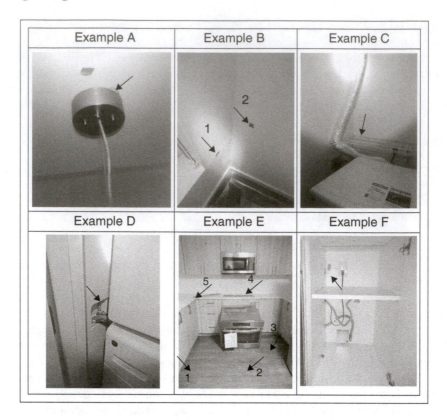

Figure 9.3 Examples of Pre-Punch Issues

review of the issue, it is evident that the cabinet faceplate did not allow the installation of the oven. Therefore, the casework subcontractor should address this issue by ordering and installing the correct faceplate; then, the appliance subcontractor can complete its installation. Finally, the flooring and countertop subcontractors can address the trade damage.

Examples B and E provide a good opportunity to remember that a project engineer does not only need to identify quality issues, but he or she is often required to review the issues within the work sequence and to identify the best approach toward their resolution. If a project engineer had simply marked the issues in Example E and requested their resolution within a week by all subcontractors without identifying the best sequence, the flooring and countertop subcontractors may have addressed their issues first. Then, the appliance subcontractor may have re-tried to install the oven and re-damaged the floor and countertop without being able to complete the task. All of these events would have happened while the casework subcontractor was still unaware that he or she needed to order the correct cabinet faceplate.

As the project nears completion, the GC will schedule the punch list inspection with the owner and/or its representatives. Generally, this official inspection is conducted by the superintendent, an owner's representative, and a representative of the design firm. The project engineer may also participate to record findings. All deficiencies are noted on the official punch list, which is a list of items remaining to be completed or corrected. If pre-punch inspections have been performed, the size of the punch list should be minimal.

When all deficiencies listed on the punch list have been corrected, the general contractor schedules the final inspection to verify that all required work has been completed in conformance with contract requirements. The final inspection is also conducted by the superintendent, an owner's representative, and a representative of the design firm. The punch list and other aspects of project close-out are also discussed in Chapter 14.

Quality control reporting

Project owners commonly require formal QC reports from the GC and expect the GC to pro-actively manage them. These reports list any deficiencies that have been identified as a part of the ongoing testing and inspection process. All deficiencies identified should be listed on a deficiency list, non-conformance-report (NCR), or QC log such as the one shown in Figure 9.4. The super-intendent will utilize the NCR log and discuss progress on open items during the weekly owner–contractor coordination meeting, which will be described in Chapter 11.

The superintendent, foreman, or quality control inspector will walk the project when each sub-contractor finishes its portion of the work to identify necessary rework. Any deficiencies noted should be listed in the NCR or QC log. As previously stated, some contractors refer to this pro-cess as a "pre-punch" or "contractor punch." All rework should be completed before follow-on subcontractors are allowed to start work. Deficiencies should be corrected promptly to minimize the size of the official punch list.

Quality control inspectors prepare daily reports, similar to the one illustrated in Figure 9.5, to docu-ment their inspection activities and any deficiencies identified. These daily inspection reports should be reviewed by the superintendent, field engineer, and all of the project management team. Non-conforming issues should be entered in the deficiency log and tracked until they have been corrected. Third-party testing agency reports are completed daily and sent to the client, designer, and general contractor's superintendent. Sometimes the city or AHJ will also receive structural inspection reports. Deviations documented in these reports should also be added to the QC log.

Reliable Construction Company
QUALITY CONTROL LOG

PROJECT: Rose Collective, Job 414 SUPERINTENDENT: Susan Thompson DATE: Aug 1, 2019
PAGE: 5 of 10

No.	Rework Description	Location	Contractor	Date Entered	Sub Corrected	GC Verified
111	Vinyl floor peeling	Unit 332	Wilson Floors	10/24/19		
112	Vandal broke storefront window	SW corner, street level	Greater Glazing	10/24/19		
113	Roof leak near stair tower	Grid C x 10	Rick's Roofing	10/24/19	10/31/19	ST
114	Concrete rock pockets in garage cols	Lower floor columns	Reliable CC	10/24/19	11/1/19	ST
115	Elevator call button malfunction	Cab #2	General Elevator	10/24/19		
116	Dead rhododendron on roof	Roof garden, NE	Green Thumb	10/24/19	10/26/19	ST

Figure 9.4 Quality Control Log

Reliable Construction Company
1401 Waterfront Avenue
Portland, OR 97210

IN-PROCESS QUALITY CONTROL INSPECTION REPORT

Date: August 15, 2019

Project: Rose Collective MXD Apartments, Job 414

Weather: Sunny, 72 degrees

Work Inspected: Green Roof System
1) Acceptability of roofing underlayment
2) Roof paver laydown – flatness, tight joints, direction
3) Topsoil sieve analysis
4) Irrigation piping: Leaks? Test pressure?
5) Size and quantity of plants
6) Subcontractor tie-off and spotter for safety

Deficiencies Noted:
1) There are a few dead plants already, especially the rhododendrons
2) There is a ponding near the stair shaft, not sure if this is a green roof or a roof issue

Inspector(s):   David Arnold/QC, and Jennifer Forsythe/PE

Figure 9.5 QC Inspection Report

Use of technology

Quality control and reporting is one of the field activities that has been most affected by new technologies. In the optics of proactive quality management, laser scanning can be used jointly with three-dimensional (3D) modeling during pre-construction to verify how the designed product will fit within the contextual conditions. Therefore, a new retaining wall between a city park and a new municipal facility can be modeled and the model inserted into the actual laser scanning of the slope. Similarly, laser scanning is used to compare the as-built product against the designed product that was modeled in 3D. When measurements are not needed, laser scanning is replaced by 360-degree photography. For instance, a contractor working on an accelerated schedule could quickly perform pre-cover surveys of the interior of a building through 360-degree photos. This task used to be performed through standard digital photography, but 360-degree photography has speeded up its performance. For instance, a contractor needed six to nine hours to perform a traditional pre-cover photographic survey of 11 apartments, whereas they can now capture the same surface through 360-degree photography in one to two hours, which is a consistent saving.

Drones are another technology used in field activities. Some construction firms utilize them to verify the quality of elements that are not easy to reach. For instance, a drone equipped with a high-resolution camera can fly around a high-rise building and allow an inspector to visually verify the quality of caulking on the envelope. Tablet apps are also prevalent in the field. For instance, Autodesk BIM 360 Field and Procore are used to run quality checklists, record issues during pre-punch and punch list inspections and communicate issues directly to the party responsible for addressing it. These tools are designed to enhance the work of project engineers, not to replace a good understanding of built environment processes. Costly errors may be created by young project engineers who over-rely on these technologies without using them within the context of their education and field knowledge.

Project engineering applications

Implementation of field construction QC is ultimately the responsibility of the superintendent, but project engineers in the field support this process in a variety of fashions:

- Process timely and specification-compliant submittals,
- Receive material deliveries and verify that the product matches the approved submittals,
- Support direct crafts and subcontractors by answering their questions and/or processing detailed and timely RFIs through the design team,
- Logging and routing testing and inspection reports,
- Logging NCRs and expediting their resolution,
- Witnessing and documenting successful MEP start-up, testing, and balancing,
- Participating with pre-punch and punch list walkthroughs, including possibly authoring the formal punch list,
- Expediting subcontractor resolution of punch list issues and punch list log updating, and/or
- Actively participating in close-out activities (see Chapter 14).

Summary

A proactive quality control program is essential to project success, both in terms of a satisfied client and a profitable project for the general contractor. Poor quality work costs the contractor both time and money and can result in the loss of future projects from the client. Because subcontractors play such a major role in construction, only best value subcontractors should be employed. Subcontractor selection based only on the lowest bid will not necessarily assure the highest quality, possibly just the opposite. The project manager, project engineer, and superintendent must ensure that all materials and work conform to contract requirements. Quality materials must be procured, and qualified craftsmen and subcontractors selected to install them. Workmanship must meet or exceed contractual requirements. An effective quality management program achieves quality standards with a minimum of rework, increases cost efficiency, improves the likelihood of schedule compliance, and satisfies the client and design team.

QC inspectors should inspect all work, whether performed by the GC's direct workforce or by subcontractors. Materials are inspected for conformance with contract specifications upon arrival on the jobsite. Mock-ups may be required to establish workmanship standards. Each phase of work is inspected as it progresses, and a punch list inspection is conducted when most of the work has been completed. All deficiencies identified during this walkthrough are recorded on the punch list. The final verification inspection is conducted by the design team after all outstanding deficiencies have been corrected.

Technology is changing the way project engineers perform quality control and reporting, but it does not replace the basic knowledge requirements to operate successfully in construction. Useful technologies can result in costly errors if this cardinal rule is forgotten and project engineers use technology tools as black boxes without verifying how their use may enhance or contrast with the objective of delivering a quality product.

Review questions

1. Why do clients, rather than the GC, contract with third-party testing agencies?
2. Why are materials that are stored on site periodically inspected?
3. Why should subcontractors not be allowed to build over work that was improperly done by a previous subcontractor?

4. What is the primary objective of an active quality control program?
5. What processes are used to actively manage quality on a construction project?
6. Why are project-specific quality control plans needed?
7. Why are mock-ups required?
8. Why should quality control inspectors be involved in the submittal process?

Exercises

1. What quality control organization would you recommend the contractor use on our civil case study project?
2. In addition to the apartment unit, what are two mock-ups that the general contractor should require subcontractors to construct for the Rose mixed-use apartment project?
3. Develop a quality control checklist for the storm water system, including catch basins (or another significant assembly) for our civil case study.
4. What might be two additional tasks we could add to the QC plan task list shown in the Project-specific quality control planning section above?

Section C
Office engineering

10 Design review

Introduction

In this chapter, we will discuss contractor involvement with reviewing design documents, during both pre-construction and construction. During the pre-construction phase of a construction project, design review involves value engineering studies and contractor and consultant proposals. The design team collects design corrections and change revisions from various sources and prepares pre-bid addenda. During construction, contractors further the design review with design submittals, such as shop drawings and requests for information (RFI). The process of incorporating changes into the design is a collaborative effort by both the design and contractor teams.

Design review during pre-construction

The pre-construction phase happens before construction occurs, of course. There are many activities performed during pre-construction by the project owner and by the design team. On design–bid–build (DBB) projects, pre-construction activities are owner-led, and contractors have limited to no involvement with the project owner and design team during design development.

However, the emergence of novel project delivery methods is providing the general contractor (GC) with opportunities to participate in the pre-construction phase. On construction management at risk (CMR) projects, owners will engage a general contractor early in the design process for a fee to assist the designer and owner with the quality of their design. Under CMR, design activities are still owner-led. On design–build (DB) projects, the design-builder entity—usually led by a GC—will also be secured early in the design process. However, different DB models will result in a different governance model for the design phase. Under progressive DB, the owner will lead or co-lead pre-construction activities and the GC will provide assistance, similarly to CMR. Under bridging DB, the owner will lead pre-construction activities and the GCs will review the outcome while preparing their proposals and negotiating the contract. Under competitive DB, the design-builder will review what has been done by the owner up to the request for proposal (RFP) and will lead pre-construction activities afterward.

Some of the activities that will be performed during pre-construction include:

- Constructability review, including evaluation of potential changes to the design to make it easier to build (i.e., less cost and time) and with improved quality and safety goals in mind,
- Budgeting,
- Scheduling,
- Identification of long-lead materials (possibly bidding and awarding) and ordering long-lead materials and equipment (such as elevators and structural steel), and
- Value engineering.

Some public agencies may employ an independent pre-construction services team in lieu of a GC on a consulting basis; this team will contribute to many of the above activities as well.

Value engineering

Formal value engineering (VE) is a methodical process to evaluate a project's design in order to obtain the best value for the cost of construction. This includes analyzing selected building systems. The goal of such analysis is to seek cost-efficient methods of performing the same function as the original design but at a lower life-cycle cost and without sacrificing reliability, performance, or maintainability. Value engineering studies may be performed at all stages of project delivery and by different project participants. However, the effectiveness of VE studies decreases as the design advances. The most effective time to conduct these studies and to significantly impact costs is during early design development, as was discussed in Chapter 2 and shown in Figure 2.4.

Independently from when they occur, the goal of value engineering studies is to select the highest value materials and systems for construction. The essential functions of each component or system are studied to estimate the potential for value improvement. The VE study team needs to understand the rationale used by the designer and their team in developing the design and the assumptions made in establishing design criteria and selecting materials and equipment. The intent of value engineering is first to develop a long list of alternative materials or components that might be used, often through a brain- storming process. Preliminary cost data are designer generated, and functional comparisons are made between the alternatives and the design components being studied. The goal of the team is to determine which alternatives may meet the owner's functional requirements and provide more value to the completed project. To this end, estimated life-cycle cost data should be developed for each alternative. The advantages and disadvantages of each VE alternative are identified, and a shortlist of options—representing those with the best values—is selected for refinement in preparation for presentation to the designer and the owner.

There are a series of scenarios for project parties to implement value engineering at different stages of a built environment (BE) project:

1.   The project owner and their consultants before a contractor is brought on a project,
2.   The contractors competing for the award of a construction contract,
3.   A construction firm during design development as a pre-construction service, and
4.   The general contractor during construction.

The *first scenario* will not be discussed because it does not involve a construction firm and their project engineers. The *second scenario* occurs when the procurement process promotes value engineering by incorporating it into the selection process. Under this scenario, contractors are invited to propose value engineering ideas to that owner. If approved, these ideas will enhance the scoring of the proposing contractor. This approach is easy to adopt in private sector procurement. However, its implementation is more difficult in the public sector, where contract award must be transparent. Still, in the predominantly public heavy-civil sector, this approach is gaining popularity by allowing proposers to submit *alternative technical concepts* (ATCs) to the project owner for their approval. ATCs are deviations from the project concept, design, or specifications. Once approved, an ATC can be incorporated into the final proposal, which provides a competitive advantage for the contractor proposing it. For instance, a contractor competing on a design–build contract for a highway interchange may submit an ATC to change the interchange schematic diagram, which would save the owner a substantial sum in land acquisition and environmental mitigations while providing an infrastructure that performs equally as well as or better than the original design. If the owner accepts this suggestion, this contractor will be able to incorporate the new diagram in its proposal and it will be scored higher thanks to the lower price proposal. ATCs are either design-centered or construction-centered. The former are usually proposed by the design consultants on the contractor team and priced by senior project engineers or estimators under the project manager's supervision. The latter are proposed by the superintendent and other senior site management and priced by senior project engineers or estimators under the project manager's supervision.

Under the *third scenario*, the contractor is brought on board before design completion, so it may review and price the design to date while supporting the owner-led team during the completion of the design. This scenario mostly occurs under CMR and DB delivery methods. The *fourth scenario* always occurs when a construction contract is in place, and the GC continues VE throughout construction to promote internal efficiency and maximize profit. However, some contracts incentivize the GC into submitting VE change proposals for approval by the project owner by sharing some of the resulting savings.

Under the *fourth scenario*, the next step in the VE process is preparation of formal value engineering proposals. The VE proposal looks very similar to a post-contract change order, including all detailed costs, markups, and substantiation. Detailed technical and cost data developed to support the recommendations should be included. The advantages and disadvantages of each option are explained. Each VE proposal is tracked in a log, similar to other document control or tracking logs managed by the GC's project engineer (PE), including requests for information (RFIs), submittals, change orders, and close-out logs. The VE proposals are submitted to the designer and the owner for approval. If approved, the proposals are incorporated into the design. If this occurs during the bid or negotiation process, the results may be issued through addenda, as discussed below. Value engineering proposals approved after the construction contract is awarded must be incorporated into the contract with a formal contract change order. When project owners want to incentivize post-award submission of VE proposals, they may include a separate contract clause that distinguishes change order proposals from *value engineering change proposals* (VECPs). Under the VECP clause, the contract describes how VE cost savings will be shared among the parties. In any case, changes will need to be estimated. Change order estimates are introduced in Chapter 20.

Addenda

Addenda (plural for addendum) are formal revisions to bid packages originated by the design team. There are a variety of reasons for the issuance of addenda, including:

- Added scope from the designer and/or project owner,
- Correction of design errors discovered either by the designer or by contractors while preparing their estimates,
- Incomplete bid package from the design team,
- Incorporation of meeting notes documenting contractor attendance at the pre-bid meeting, as well as contractor-generated questions and design and owner team answers, and
- City permit comments and revisions.

There is no exact quantity of addenda required to be issued. Some projects will receive only one addendum, others several. Language will be included with each addendum requiring contractors to notify all subcontractors and suppliers of the content. It is customary that addenda are not issued during the final week of bid preparation, but occasionally contractors will receive the last addendum as late as the day the bid is due, which makes it extremely difficult to incorporate it into their estimates. The GC will be required to note on its bid form all of the addenda that they received and incorporated into their bid. An omission of an addendum from the bid form would be sufficient reason to reject that contractor's bid.

After the contractor is selected, the design team should incorporate all of the addenda and issue the final set of contract drawings. It is easier for the builders to construct the project from one comprehensive set of drawings and specifications, rather than a multitude of addenda. We will discuss the handling of document records, including versioning of drawings and specifications, in Chapter 11.

Design review throughout construction

A design package is rarely 100% complete at the time of the contractor's bid or even when construction commences. Even when construction expertise is available during pre-construction, it is expected that the design will further develop during construction, especially early on, but at a slower pace than during pre-construction. Many elements of the design are too detailed to include with the bid documents, and many others are left up to the contractor's means and methods. This is true especially with subcontractors that have a design–build scope, such as the mechanical, electrical, and plumbing (MEP) contractors. The routing of one-inch electrical conduit and one-half-inch copper pipe would be difficult for the design team to accurately draw and dimension. Components that require off-site fabrication, such as structural steel and casework, are other scope items that continue to be designed throughout the early phases of construction. The design team reflects on their drawings where the final elements are to be placed, but their exact detail and sizing is often left up to the supplier or fabricator.

Specialty contractors and suppliers submit their portion of the design to the GC, the design team, and the project owner for their reviews. Design submittals act as the contractual vehicle to receive preliminary acceptance before further financial commitment. However, submittals are not limited to design deliverables but are also used to seek preliminary approval to purchase materials, fabricate assemblies, or install materials or assemblies. We will discuss all the different types of submittals in Chapter 11. In this section, we will elaborate on one major type of submittal, design submittals (such as shop drawings), which reflect one of the contractors' major contributions to evolving design during construction. In addition, we will introduce requests for information. RFIs are used by contractors to ask questions and clarify the design team's intent throughout the course of construction. In this chapter, we will discuss, in detail, how responses to RFIs are originated. In Chapter 11, we will discuss how responses to RFIs are recorded and documented on the contractual documents.

Design submittals

Contractors prepare design submittals to reflect their interpretation of the designer's intent and send those submittals to the designer for confirmation and, hopefully, approval. One of the major types of submittals originated by the construction team is shop drawings. Many in the built environment industry refer to all drawings prepared by second-tier vendors as shop drawings, but that is not technically correct. In fact, design submittals include:

- Shop drawings,
- Fabrication drawings,
- Installation drawings, and
- Coordination drawings.

The processes of creating and processing different types of design submittals are similar. The specifications will indicate which areas of work require design submittals, such as shop drawings. To prepare these design submittals, it is customary for the supplier or subcontractor to employ design engineers, who will prepare these drawings for approval, before proceeding with construction activities. Before submitting them to the owner and their consultants, the GC is required to review these design submittals to verify that they meet the intent of the specifications and the dimensions are accurate. Figure 10.1a shows an example review of a design submittal by a GC. These design submittals are sometimes used by subcontractors to inquire about design uncertainties. Therefore, the GC's project engineer will also need to verify if these inquiries are legitimate and would need the submission of an RFI or confirmation that the information is available somewhere in the contractual documents. Figure 10.1b shows an example of such an inquiry.

These drawings are created in a larger scale and will include sizes and details that the contract drawings did not contain. Some of the materials and systems that are customarily submitted as a design submittal include:

- Shoring systems,
- Concrete reinforcement steel,
- Structural steel,
- Wood and steel trusses,
- Fireproofing,
- Curtain-wall and storefront systems,
- Elevators, and
- Mechanical, electrical, and plumbing systems, including materials and equipment.

Requests for information

Requests for information are one of the most important tools for contractors during the construction control and administration phase. RFIs allow contractors to submit inquiries to the designer and the owner on unclear aspects of the design. Therefore, RFIs help to bring the design to the level of definition that is necessary for quality, quick, and efficient construction. Even with the technology tools available today, some conflicts will not be discovered until work is well underway. Many unforeseen conditions are exposed during construction, such as when the backhoe excavates for the site utilities or during selective demolition. Similarly, on building projects, the contractor may discover that the fire protection heads and lights are roughed-in for the same locations or the light pendant fixtures on the kitchen countertop island will interfere with the cabinet doors. RFIs are a specific type of written communication that can be used to document questions and record responses on such issues.

If any member of the construction team discovers a construction document or site condition discrepancy and needs clarification from the designer or owner for resolution, they will typically generate an RFI. Similarly to design submittals, RFIs may be transmitted on paper or electronically. RFIs can be originated by several of the construction participants; they may be written either by subcontractors, by suppliers, or by the general contractor. Questions generated by subcontractors or suppliers are submitted to the GC's project engineer (who manages those subcontractors or suppliers) for review prior to their submission to the owner or designer. As shown in Figure 10.1B, subcontractors may embed their RFI in the design submittal, but subcontractors can also prepare a separate RFI.

RFIs are generally managed by project engineers who maintain a log indicating the status of each question, such as that shown in Table 10.1. As with most construction logs, including submittal and sketch logs, the RFI log is generally maintained by project engineers. It is important that the

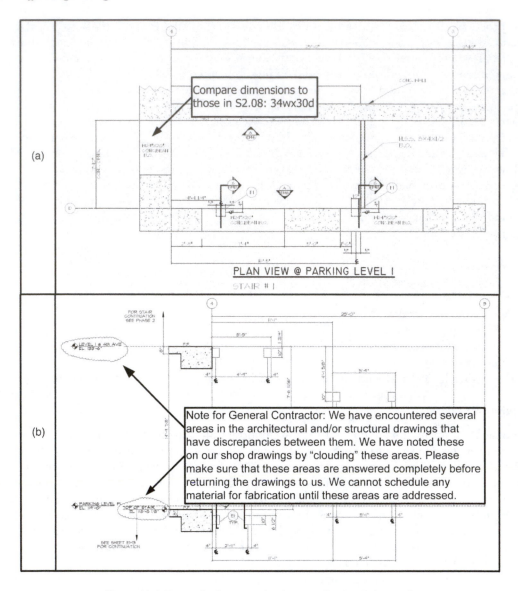

Figure 10.1 Example Communications on Design Submittals

log identifies who generated the question and who is expected to respond. Open items on the log should be reviewed during the coordination meetings with the project owner and the designer. This log and any other documents distributed in the meeting should be an attachment to the meeting notes. Meeting notes and other construction management tools will be discussed in our next chapter on document control.

Verbal responses to RFIs should not be accepted by the construction team. If the designer does not provide a written answer, the project engineer should write the designer's verbal direction in the answer portion, document the source and time, and route the RFI back through the designer. All responded and resolved RFIs should be routed to the appropriate GC office and field staff, as well as the originating subcontractor. They should also be sent to any other subcontractors and

Table 10.1 RFI Log

RFI#	Title	Initiated By	Sub RFI#	Date Sent	Assigned To	Date Due	Date Returned	Drawing/ Spec Ref	Posted to Contractual Documents	Status
001	Underpinning pile installation sequence	Sub 1	01	3/31/2015	Geotechnical consultant	4/7/2015	4/17/2015		5/28/2015	Closed
002	Drywall shaftwall clarifications	GC		4/6/2015	Geotechnical consultant	4/13/2015	4/6/2015	A6.00-A6.20	4/6/2015	Closed
003	Unit demizing wall clarification	GC		4/6/2015	Architect	4/13/2015	4/6/2015	A2.70-A2.85	4/6/2015	Closed
004	Roof patio framing	GC		4/6/2015	Architect	4/13/2015	4/6/2015	A8.26	4/6/2015	Closed
005	Shoring tiebacks conflict with side sewers	GC		4/9/2015	Geotechnical consultant	4/16/2015	4/23/2015	SH drawings	5/29/2015	Closed
006	Garage entry door	GC		4/16/2015	Architect	4/23/2015	4/17/2015		4/16/2015	Closed
007	Crane surcharge W-2 to W-5	Sub 1	02	4/27/2015	Structural engineer	5/4/2015	4/28/2015		5/29/2015	Closed
008	Utility sheds pile interference at north property line	GC		4/29/2015	Owner	5/1/2015	4/30/2015	SH2.0	5/29/2015	Closed
008.1	Underpinning piles at utility sheds	GC		5/12/2015	Geotechnical consultant	5/13/2015				Open
009	Soldier pile utility conflicts at north-east corner	GC		4/29/2015	Geotechnical consultant	5/1/2015		SH2.0		Open
010	Lagging attachment to south-east property footing	Sub 1	03	5/4/2015	Geotechnical consultant	5/5/2015	5/5/2015		5/29/2015	Closed
011	Underpinning benching	Sub 1	04	5/4/2015	Geotechnical consultant	5/5/2015	5/6/2015		5/29/2015	Closed
012	Pile S–1 to S–4 modifications	Sub 1	05	5/5/2015	Geotechnical consultant	5/5/2015	5/5/2015	SH3.2	5/29/2015	Closed
012.1	Pile S–1 to S–7 modifications revised	Sub 1	06	5/6/2015	Geotechnical consultant	5/8/2015	5/7/2015	SH3.2	5/29/2015	Closed
013	Stud rail base plate size	GC		5/5/2015	Structural engineer	5/12/2015	5/14/2015	S7.06	5/29/2015	Closed
014	South building foundation over property line	GC		6/26/2015	Structural engineer	7/3/2015				Open
015	Existing conditions – 8" side sewer	GC		5/7/2015	Architect	5/14/2015		C6.00		Open
016	Foundation at pile S–4	GC		5/18/2015	Geotechnical consultant	5/25/2015	5/18/2015		5/29/2015	Closed
017	Tower crane foundation pier move	GC		5/18/2015	Structural engineer	5/25/2015				Open
018	Piles N-1 to N-8 modifications	Sub 1	07	5/19/2015	Geotechnical consultant	5/21/2015	5/19/2015		5/29/2015	Closed
019	Duct bank routing in garage	GC		5/26/2015	Architect	6/2/2015				Open
019.1	Duct bank routing in garage – Revised	GC		7/27/2015	Architect	8/3/2015	8/11/2015	S2.07	8/14/2015	Closed
020	Pile N-8 modification	Sub 1	08	5/27/2015	Geotechnical consultant	5/28/2015	5/27/2015		5/29/2015	Closed
021	P-7 plumbing sump vaults – Mat footing	Sub 2	01	6/8/2015	Structural engineer	6/15/2015	6/15/2015	S3.01	8/14/2015	Closed

suppliers who may be involved or impacted. All answered RFIs should be reviewed for impacts on the project scope and schedule.

Regardless of the author or originator of the RFI, there are a few basic rules that should be followed by project teams.

- *RFIs should be submitted in writing*: Asking questions and receiving responses is only one part of proper RFI management. It is crucial to make sure the response is available to everyone who needs it. On a typical building or civil construction project, hundreds of RFIs may be needed, so tracking them properly is important to ensure a response has been received and the people needing that response have been informed. Project engineers may contact one of the owner's consultants with time-sensitive questions, so these verbal questions should be followed up with written confirmation in the form of an RFI with a proper tracking number.
- *Subcontractors and suppliers must route RFIs through the GC*: The GC is ultimately responsible for managing RFIs, verifying their validity, and updating contractual documents with RFI responses, as we will discuss in Chapter 11. Many subcontractors will want to use their own RFI forms and not the GC's or the project owner's. If the GC allows this, these subcontractors' RFIs will then be submitted as an attachment by using an official cover with the RFI tracking number. Maintaining consistency in RFI tracking numbers is crucial for documenting responses on contractual documents.
- *RFIs should be submitted according to the requirements in the contractual documents*: Most general contractors or subcontractors would prefer to use their own form, but some contract specifications will dictate the form to be used, especially on public works lump sum or unit price projects. Even on private projects, some architects or project owners may have their own customized forms that they wish the contractors to use, which will have been identified in the contract special conditions.
- *RFIs should be focused*: The number of separate clarifications requested on a single RFI should be limited. If too many questions are asked, the designer may wait to respond until he or she has answers to all of the questions. This may delay the overall response. On the other hand, if several issues are related to each other, it may be best to consolidate them under an individual RFI, rather than submitting piecemeal RFIs that may result in inconsistent responses.
- *RFIs should be clearly written*: In order to elicit a clear answer, each RFI must be written in an objective manner and be complete, concise, and accurate. It must reference specific contract and specification sections and should be supported with appropriate documentation. Supporting documentation may include specification sections, photographs, drawings, and sketches.
- *RFIs should lead to a buildable response*: When formulating questions on an RFI, it is important to convey what responses are viable for construction. If the RFI is asked too broadly, it may result in responses that are not buildable or may conflict with other contractual documents, schedule constraints, or price assumptions. When the risk of conflicts arises, it is best for the project engineer in charge to have the RFI reviewed by the superintendent or project manager before submitting it.

Although RFIs often carry a negative connotation, a proper RFI process can be a valuable tool for the construction team. Unfortunately, some contractors looking for change order opportunities may generate unnecessary RFIs. As a result, some designers and project owners dislike the RFI process, because it costs them time and money to respond. This negates the RFI's role as a design-continuation tool. If a conflict is discovered early enough, the RFI can actually mitigate a change order, and in that way, it is a contractor's contribution toward continuation of the design process and helps with quality control. RFIs are an important early step in the quality control process. The quantity of RFIs generated on any project depends upon the design and construction teams, the contract format, the complexity and value of the project, and the quality of the design documents. The more complete and coordinated the design, the fewer RFIs will arise in the field. Project teams that have worked together previously tend to have fewer questions. Also, the project delivery method and contract type are known to affect the number of RFIs on a project. Usually, integrated project delivery (IPD) and DB projects tend to have significantly fewer RFIs than DBB. Also, CMR delivery usually leads to a lower number of RFIs. Moreover, cost plus contracts also tend to have fewer RFIs than do competitively bid lump sum or unit price projects. A sample RFI for the Rose mixed-use development (MXD) case study project is included as Figure 10.2.

RELIABLE CONSTRUCTION COMPANY
1401 Waterfront Avenue, Portland, OR 97201
(503) 719-9874

REQUEST FOR INFORMATION

Project: *Rose Collective Mixed Use Apartment*   Date: *July 19, 2018*
Area/System: RFI No.   *101*
To: *Creative Spaces, Inc*   Related RFIs: *NA*
Address: *444 Hill Street*
                Portland, OR
Attention: *Robert Jackson*
Required Response Date: *July 27, 2018*

Subcontractor/Supplier Forwarding Question: *NA*   Sub's RFI No. *NA*

Subject: Door 6-06 Fire Rating

Detailed Description/Request:

The door schedule calls out the 6-06 conference room door as a 20 minute rated door. However, this door has to be installed within a one hour rated wall. Please confirm the fire rating of this door.

Attached/Referenced Drawings/Photographs/Specifications:

Cost or Schedule Impacts: *Unknown at this time*

Please reply to: *Jennifer Forsythe, Project Engineer*

Architect/Engineer Response:

Signed: *Robert Jackson*   Date: *July 22, 2018*

File Code: *414/RFI*

Figure 10.2 Example RFI

Incorporation of design changes

Before bids are submitted, design changes are documented through the use of addenda, as discussed above. After the contract has been executed, design changes are incorporated through formal contract change orders, as will be discussed later in Chapter 20.

During construction, design changes may be originated in a variety of fashions, including answers to requests for information, as just discussed. Approved copies of design submittals and shop drawings may also include changes in the design. In addition, there are a variety of formal documents that the design team may utilize to inform the construction team of changes, including:

- Sketches (SKs);
- Architect's supplemental instructions (ASIs), for example, American Institute of Architects (AIA) form G710;
- Field orders or field work orders (FOs or FWOs);
- Construction change directives (CCDs), for example: AIA form G714 (see Chapter 20);
- Construction change authorizations (CCAs);
- Work changes proposal requests (PRs), for example, AIA form G709;
- Meeting notes, which may also document a change in direction from the owner or designer; and
- New or revised full-size drawings or complete specification sections.

Contractors will document all design-changing documents and track them on logs. Many of these documents will also be reviewed at the weekly meeting with the owner and designer. Ideally, any design change will be incorporated into the contract documents, including full-size drawings and specification sections, and will be issued to the GC in a request for change order or a change order proposal (COP). However, some project teams find it more efficient and economical to work from these documents without issuing a new set of drawings or specifications. Incorporation of these changes into the contractor's set of record drawings is discussed in Chapter 11.

Use of technology

Several technologies are changing the way contractors review design, including the use of electronic plan rooms, building information modeling (BIM), and laser scanning.

Construction projects have previously relied on the use of paper-based plan rooms as a central repository for all construction documents, including full-sized drawings and specifications. The downside of this approach is that it is difficult to keep all documents up to date with the most recent information, including change orders and RFI responses. Moreover, to access information, a project team member needs to be in the room. To overcome these issues, contractors are increasingly using the concept of electronic plan rooms, in that all project documents are in an electronic repository, either a project server or a remote cloud computing repository. Although physical plan rooms may still be implemented, they are enhanced through touchscreen displays, mobile computing stations, and tablets to allow project members to access the latest set of documents. Various software applications are used to support the implementation of electronic plan rooms, including Bluebeam, BIM360 docs, and PlanGrid.

Today, the use of building information modeling tools is pervasive in the construction industry, especially in the building sector. These tools will be discussed in detail as part of Chapter 15, but it is worth mentioning here how BIM fly-through simulations are used to perform cross-reviews of different design packages. Fly-throughs allow project engineers to virtually implement all design packages and perform clash detection analyses to identify conflicts and inconsistencies.

For example, a conflict between electrical conduits and mechanical ducts could be detected by modeling the electrical and mechanical shop drawings and running clash detection.

Another technology that is frequently used to verify site conditions is laser scanning. If a contractor has some doubts on the site conditions reported on the drawings, they could use laser scanning to survey the existing conditions and compare them against the assumed conditions in the drawings.

Project engineering applications

Pre-construction services are not typically performed by project engineers, other than assisting chief estimators or project managers. A project engineer may become involved during pre-construction by soliciting and expediting subcontractor value engineering proposals and tracking them on the VE log. Routing of addenda to every subcontractor and supplier who might be bidding on a project is difficult, but the project engineer can make sure that known subcontractors and suppliers have been notified. During construction, the project engineer will also manage shop drawing logs, including expediting and reviewing all submittals. Project engineers play a major role in the management of requests for information. The GC's own foremen will bring in their questions. The project engineer will review the contract documents and, if answers are not readily available, will prepare RFIs on behalf of the foremen. Subcontractors prepare requests for information and submit them to the general contractor's project engineer for review, who logs and forwards them to the design team. Management of the RFI log and discussion of outstanding issues at the owner–designer–contractor meeting is a project engineering responsibility. There are many methods that the design team may use to change the design, and it will also be the project engineer's responsibility to log and disseminate these documents to subcontractors and suppliers and eventually incorporate them into the GC's set of record drawings.

Summary

General contractors who have an opportunity to participate in design review during the pre-construction phase can help the project owner and designer seek alternative methods and materials that will provide better value to the project. Value engineering proposals are the vehicle that project managers and estimators use to research these topics and provide specific cost and scheduling impacts. Any questions or changes that affect the contract documents before the project has received bids must be documented by the design team and issued to bidding contractors through formal addenda. All addenda are typically incorporated into the documents after bid, and a "for-construction" contract set of drawings and specifications is produced.

The construction team continues to input to the design throughout the construction phase, especially in the early stages. Contractors prepare shop drawings and other forms of submittals for areas of work that were either too detailed for the design team to effectively draw or that required off-site pre-fabrication, such as structural steel. Most projects will have some issues that require additional clarification from the designers by the construction team. Requests for information are the most common construction management tool used for this process and are an active method of quality control. Some projects, especially competitively bid public works jobs, can experience thousands of RFIs, so successful management of that process is critical to the success of the project. If discrepancies are discovered early enough, RFIs can actually reduce the quantity of change orders.

It is important to follow contractually prescribed procedures to ensure a successful RFI process, especially on projects that have many questions. All contractors should be clear and accurate in

RFIs. The specific problem must be stated in simple and objective language. Photocopied portions of drawings or specifications should be attached, if needed to clarify the issue. Photographs of exact field conditions are also often attached. Test reports, letters, submittals, or other documentation may also be needed. Even if the reviewing party already has some of this information, the question must be drafted so that it is completely self-explanatory. The RFI author should do all that is possible to assist the designer in answering the question. All pertinent information must be included. If the project engineer knows of a good and logical solution, he or she should recommend it. However, the contractor should not assume how the architect or engineer would respond and not forward the question nor should they answer it themselves and be in error.

The GC's project engineer should diligently review each RFI before forwarding to the designer. During construction, many designers have often moved on to another project and/or their construction administration budget may be nearly expended. The construction team should not waste the designer's time and money, challenge their patience, and jeopardize the team relationship that has been built by asking irrelevant or superfluous questions. The construction team should attempt to find an answer in the documents internally before submitting the RFI to the designer. But, if the answer is not clearly in the documents, the question should be processed. All design changes must be incorporated into the contract documents. There are a variety of tools used by both the design and construction teams to accomplish this, including technology tools.

Review questions

1. Before a project has been bid, designers issue addenda changing the design. What do they issue post-bid?
2. Are all value engineering proposals cost reductions? Why might a project owner consider a change that increases the price?
3. Why does the design team not indicate the exact location of each item in the project?
4. Why should the GC diligently review a vendor's shop drawings?
5. Why are RFIs used on construction projects?
6. Why do some designers resist RFIs?
7. Why do some designers welcome RFIs?
8. When is the RFI log reviewed?
9. Who should prepare an RFI?
10. Are verbal RFI responses sometimes acceptable? If so, how should they be ultimately documented?
11. Do all RFIs result in change orders?

Exercises

1. You have been asked to conduct a value engineering study to determine the most cost-effective type of curtain-wall window system for the Rose case study project retail level. What functions would you consider essential in your analysis and what cost data would you consider in conducting the study?
2. Today it is customary for the designer to require the GC to stamp all submittals with an "approved" stamp and the GC, in turn, to require the same from their vendors before submitting. Why?
3. As the GC's PE on the Rose mixed-use case study project, prepare an RFI asking for direction regarding the upgraded elevator cab finishes.

4. As the GC's PE on the I-90 overpass project, prepare an RFI where the contract documents are in conflict with the actual site conditions.
5. Assume you are a GC competitively bidding a privately owned new warehouse against seven other contractors. In the request for quotation, the client has asked for voluntary VE proposals to accompany your bid. Do you comply? Why or why not?
6. How would your answer to Exercise 5 differ if you were the only GC proposing on a negotiated project?
7. What are the advantages and disadvantages (there are several) of (a) the design team issuing all changes with new full-size drawings and complete specification sections or, conversely, (b) never revising the contract documents and only issuing 8½ × 11 design changes?

11 Project documentation

Introduction

Early on in their careers, project engineers will be more involved with document control processes than with cost, schedule, quality, and safety controls directly. Project site management staff—including project managers, superintendents, and project engineers—do not build construction projects with hand and power tools like carpenters, laborers, ironworkers, plumbers, and electricians. Rather, they make decisions and document them in paper or electronic records. Therefore, one of their tools is proper document management and control. Any discussion of construction management revolves around construction management documents; in a sense, almost every component of construction management relies on document control. As shown in Figure 2.1, all of the pillars of project success utilize documents as a means of communication. Although document control tools are discussed throughout the whole book, this chapter summarizes some of them. Some of the documents that site management use as tools include, in no particular order:

- Requests for Information (RFIs),
- Proposals,
- Architect's supplemental instructions (ASIs),
- Subcontracts,
- Purchase orders,
- Lien releases,
- Submittals,
- Payment requests,
- Schedules,
- Estimates,
- Warranties,
- Operation and maintenance (O&M) manuals,
- Prime contract agreements,
- Memorandums,
- Non-conformance reports (NCRs),
- As-built drawings,
- Shop drawings,
- Safety inspections,
- Logs,
- Letters,
- Daily job diaries,
- Change order proposals,
- Punch lists,
- Work packages,
- Cost forecasts,
- Insurance certificates,
- Permits,
- Meeting notes … and many more.

In Chapter 10, we introduced shop drawings and RFIs. In this chapter, we will integrate those topics again and discuss in detail submittals, the superintendent's daily job diary, meeting notes, and record document creation and maintenance.

Submittals

In our last chapter, we introduced design submittals, such as shop drawings prepared by contractors and suppliers, as a means of continuing the design process through the early stages of construction. In this section, we will introduce other types of submittals and discuss the submittal process.

The preparation and processing of submittals is a major element of active quality control. By definition, the *submittal* is a document or product generated by the construction team to verify that what they plan to purchase, fabricate, deliver, and ultimately install is in fact what the design team intended by their drawings and specifications. It serves as one last check of the design validity. Submittal requirements for a project are contained in the specifications of the contract. On a building project, the procedures for processing submittals are generally in CSI Divisions 00 and 01 of the contract specifications. On civil projects, submittals include drawings, test results, and any other documents, as described in the governing standard specifications. For instance, federally funded roadway and bridge projects refer to the Standard Specifications for the Construction of Roads and Bridges on Federal Highway Projects by the U.S. Department of Transportation.

Although this is not their original intent, submittals also serve as a tool for the general contractor's (GC's) project team to flush out errors that the subcontractors and suppliers may have made in their bids, especially with product substitutions. Although subcontractors and suppliers are supposed to have bid their scope according to the subcontract, which incorporates plans and specifications, the GC does not want to discover too late that they have misinterpreted the contractual guidelines.

Example

A general contractor does not want to be in the situation of receiving 100 kitchen refrigerators that lack ice-makers if the contractual documents required refrigerators equipped with this feature. Although the supplier could solve this issue by retrofitting the refrigerators in the field, the resulting two-week schedule delay will also affect the schedule of the plumber who has completed his or her rough-in for the ice-makers and is expected to connect them. If a supplier had been required to submit the product data for the refrigerators before ordering them, this issue would not have arisen. Moreover, even if the issue can be solved, the two-week delay may impede turning the building over to the project owner in a timely fashion, which will delay the leasing of the apartments. In return, the project owner will delay release of retention funds to the general contractor and may request compensation for the delay, if this is allowed by the liquidated damages clause in the prime contract.

Submittals come in a variety of formats. As stated, *shop drawings* and other *design submittals* are a major group of submittals. Early on in the construction phase or late in the pre-construction phase, construction *mock-ups* are also common for specific areas of work to allow the design team to establish expected quality levels. Mock-ups are often regarded by contractors to be expensive submittals, because that work is performed out of sequence. There are a variety of *product data sheets*, including safety data sheets (SDSs), that will be provided to the general contractor by all construction firms delivering materials to the job. Product data includes *color charts* for finished materials and physical *samples* of materials, which can be bricks, tiles, carpet, and millwork. A sample laminate material already approved by the architect is shown in Figure 11.1.

Although adequate for cost estimating by contractors and for construction of some basic scope items, contract drawings prepared by the design team typically do not show sufficient detail to be suitable for fabrication and production of many required construction materials. In addition to submittals being one of the first steps in the quality control process, they are also one of the final steps in the design process, as discussed in the last chapter. The manufacture or fabrication of construction materials often requires that the contract drawings be amplified by detailed shop drawings that supplement, enlarge, or clarify the project design. These descriptive drawings are

Figure 11.1 Submittal of a Physical Sample of Material

prepared by the manufacturers or fabricators and are provided to the purchasing organization, either the GC or a subcontractor. Shop drawings and other submittals are reviewed by the GC's project engineer to ensure conformance with the contract drawings and specifications before being forwarded to the designer. *Submittal cover sheets* are used to transmit these materials to the designer. The format for the cover sheet may be prescribed in the special conditions of the contract.

Process

Managing submittals is an important duty of a general contractor during the construction control and execution phase. To this end, a general contractor is required to identify a process to promote consistency among all the individuals working in the project team. One of the first steps for the GC's project team, including project engineers, is to skim and scan the contract specifications for submittal requirements. Ideally, this is done before awarding subcontracts and purchase orders so that the submittal expectations of the GC by the design team are passed along to the second-tier vendors. Submittal planning involves the development of a submittal schedule, which is given to the project owner and his or her consultants for review, and an expediting or submittal log to manage the submittal process with subcontractors and suppliers. Essentially, the GC will "submit" the "submittal schedule" for approval. This is to ensure that all items the owner and designer expect to be submitted will be accounted for. A flowchart of the submittal planning process is shown in Figure 11.2. In addition to identifying items that require submittal,

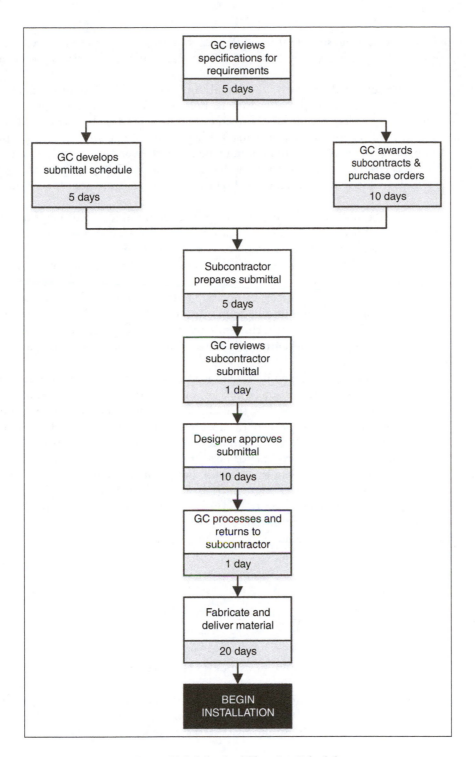

Figure 11.2 Submittal Planning Schedule

the GC should also review the submittal schedule with the project superintendent to verify that the timing of the submittal process will support off-site fabrication, material delivery, and the construction schedule.

Upon receipt, each submittal should be thoroughly reviewed by the GC to confirm compliance with the contract specifications and drawings, because vendors may attempt to substitute a less-expensive, non-specified product. This review should minimize the chances of the GC's project manager and project engineers later being in the position of explaining, for example, why the millwork supplier submitted oak trim molding when a more expensive vertical grain Douglas fir trim molding was clearly required by the contractual documents. In addition to reviewing the trim molding sample, the shop drawings should also be reviewed for dimensions that fit the current installation.

Each submittal is assigned a number and entered into the submittal log. This log is updated and discussed at the weekly owner–contractor coordination meeting, as well as at subcontractor coordination meetings. To maintain focus on what is pending approval, so as to expedite its resolution, discussion should be limited to submittal materials and documents that the project owner and his or her consultants have received. Some GCs will assign a submittal number for every item prior to processing any submittals. These numbers can be tied to the specification section numbers. For example, structural steel shop drawings may be simply "submittal number 8" for the entire project or they may be number "05-30-00-3" for the third submittal in steel specification section 05-30-00. Whichever method is chosen, each submittal should be uniquely numbered to allow efficient logging, filing, and future retrieval.

The special conditions of the contract will determine if the GC is to forward submittals for processing and approval to the architect, engineer, project owner, city, or directly to sub-consultant designers. The submittals may be reviewed by several people, either concurrently or sequentially. The GC's project manager or project engineer should set up the submittal log to be able to document the entire review process and all the stakeholders. After receipt of the processed submittal from the approving party, the general contractor needs to review the decision and log it accordingly. If changes have been made to materials by the owner or the designer on the submittal, the GC's project team, likely the project engineer (PE) and superintendent, should make a note for potential change orders. Depending on the project type and the project team, the range of decisions may vary. For instance, many owners and designers return submittals with one of the following four options checked on the designer's submittal stamp:

- Approved,
- Approved as noted,
- Revise and resubmit, and/or
- Rejected.

However, jurisdictional and/or legal implications may lead to different wording, such as the replacement of "approved" with "reviewed" to limit potential legal implications on design responsibilities. Also, the specifications may suggest different ranges of options depending on the given deliverable. For instance, the Washington State Department of Transportation (WSDOT) categorizes one of the design submittals, working drawings, into five categories depending on content:

- Type 1: These drawings must be submitted to the project owner at least seven calendar days before the work represented by the submittal begins for *its information*.

- Type 2 and 2E: These drawings must be submitted to the project owner at least 20 calendar days before the work represented by the submittal begins for *its review and comment.*
- Type 3 and 3E: These drawings must be submitted to the project owner at least 30 calendar days before the work represented by the submittal begins for *its review and approval.*

[Adapted from the WSDOT Standard Specifications for Road, Bridge, and Municipal Construction (WSDOT 2017; pp. 1–30)]

This is an example of how project owners may tie timing to expected decisions on submittals. Therefore, submittals that do not need review, comments, or approval only need to be submitted seven days before starting work to allow for timely inspections. On the other hand, submittals that need review and approval need an early submittal. Conditions for deciding what submittals will require early processing will be discussed in the next section.

When a review and/or approval is expected, the general contractor will monitor and record its receipt on the log. The submittal will then be returned to the originating subcontractor or supplier, which will then release the materials for order if it was approved. Other contractors and suppliers may need to review processed third-party submittals. For example, the electrical contractor will need to see the mechanical equipment shop drawings to verify that power is provided in the right voltage and in the right locations. It is the originating supplier or subcontractor's responsibility to also check for changes of scope made by the design team on the submittal disposition and promptly notify the GC.

Early submittals

The special conditions of the specifications may require the general contractor to prepare a *schedule of submittals* early in the mobilization or start-up phase of the project. Even if not required, this is a good tool for the GC to manage vendors. This may be in addition to the *submittal log* preparation discussed above. The reason the design team may request a schedule is to be able to schedule its in-house review efforts. The team does not want to get hit with 90% of the contractors' submittals all during the third week of the project. Submittals need to be turned in and approved before materials are installed on the project. Submittals for products needed for early phases of construction should be submitted as a part of project start-up. This is to provide time for the designers to approve them and for the materials to be delivered to the project site without delaying scheduled activities. If materials are released for order and/or delivered to the site before submittal approval, the supplier has taken the risk of rejection and removal of materials, which will have an adverse schedule impact. Submittals should be prioritized so that those materials needed on the jobsite early and those that have longer lead times are submitted early.

Example of Early Submittals

- Reinforcement steel (rebar) shop drawings,
- Concrete design mixes,
- Embedded steel (such as anchor bolts),
- Select backfill samples,
- Structural steel shop drawings,
- Elevator shop drawings, and
- Long-lead mechanical and electrical equipment.

Similarly to quality control, submittal timing is also critical to the success of the overall construction schedule. Late or rejected submittals delay material fabrication and deliveries, which also affects cost.

Superintendent's daily job diary

As introduced earlier, all the construction documents discussed in this and other chapters are important to a project's success, and this is particularly true of the superintendent's daily diary. Some construction professionals also refer to the diary as a daily journal, log, or report. The job diary is one of the most important construction management tools and jobsite records. It is viewed as contemporaneous documentation of each day's events. Additional uses include:

- Record of daily job activities,
- Change order backup,
- Back charge and claim defense and preparation, and
- Schedule progress.

An example entry in the superintendent's daily job diary for the I-90 case study project is shown in Figure 11.3. There are many different forms that are utilized in the industry, and most of them have common features. Each construction firm should select a standard form to use on all projects. Some of the important information that should be recorded in a superintendent's diary are:

- Project name and job number,
- Name of superintendent authoring the diary,
- Date and week day,
- Weather conditions, including wind speed,
- Manpower for all contractors, including the GC's direct crafts,
- Active work areas and daily progress,
- Hindrances or delays,
- Deliveries, including construction equipment,
- Visitors, including the owner, architect, and inspectors, and
- Accidents, although they will also be noted on an accident report.

Although completion of the diary may be delegated to the assistant superintendent, foreman, or project engineer, the project superintendent should fill out the diary. Because of its historical importance, the diary should be prepared by the superintendent, who is responsible for field activities. This individual will also retain a copy of the diary for his or her own record. The diary should be completed at the end of each day. Postponing this task until the next morning, or doing five of them at the end of the week, detracts from accuracy. Ideally, the diary should be handwritten in permanent ink. This helps support authenticity and originality. Of course, many superintendents are now preparing their diaries on their computers or tablets. To some extent, this negates the originality of this important historical record, so it is best if electronic diaries are printed out and hand-signed in permanent ink.

Subcontractors are usually required to complete daily diaries and provide copies to the GC. These diary entries provide the general contractor with direct evidence of subcontractor manpower and their view of daily progress. If there is a problem or restraint noted on a subcontractor's

GATEWAY CONSTRUCTION COMPANY
2201 First Avenue, Spokane, Washington 99205
(509) 642-2322

Daily Job Diary

Project: *Interstate 90 Highway Overpass*   Job Number: *1732*
Superintendent: *Randy Buckwater,* Date: *October 14, 2018,*   Day: *Monday*
Today's Weather: *Sunny, 56 degrees, Slight NW Wind at 7 MPH*

Activites Completed:
Shoring for upper deck is ongoing, should be complete by 11/1. Electrical subcontractor is installing underground electrical ductbank at the SE off-ramp. Flagging traffic each day.

Problems Encountered:
There was a passenger car accident, unrelated to our project about 10:00 am this morning. No one was injured, but the ambulance and police showed up. Traffic backed up considerably and impacted our truck and equipment access to the site. The State Patrol interviewed me and Cathy our flagger. I asked the Trooper to send a report to our home office in Spokane.

Materials and Equipment Received Today:
Formwork, Rebar, Light Tower; backhoe for electrician

Rental Equipment Returned:
Walk-behind jumping jack

On-Site Labor/Crafts/Subcontractors/Hours Worked:
Carpenters: *6 craftsmen at 8 hours*
Laborers: *2 men at 10 hours*
Ironworkers: *2 ironworkers at 8 hours*
CM Finishers: _____
Masons: _____
Electricians: *4 sparkies (2 men and 2 women) at 6 hours each*
Drywallers: _____
Plumbers: _____
HVAC Tinners: _____
Painters: _____
Other: *Forklift OE at 8 hours, Flagger @ 10 hours*

Figure 11.3 Sample Entry on the Superintendent's Job Diary for the I-90 Case Study Project

diary, the general contractor's project team should deal with it immediately. Some project owners may require copies of the general contractor's daily diaries, as well as the subcontractors' diaries. Similarly to the preparation of three-week schedules, the requirement for subcontractors to provide diaries should be included in their subcontract agreements.

Meeting notes

Construction meetings and meeting notes are also extremely valuable processes and tools on any project. Construction meetings provide an avenue for direct and timely communication and information exchange. Creating the action and obtaining the decisions necessary to maintain the scheduled flow of work enhances that communication. The meeting itself is a tool that should be used to assist in construction of a project. There are many different meeting types; some of the more common include:

- Pre-bid meetings, conducted by the lead designer;
- Pre-construction meetings:
  - between the project owner, designer, and contractor,
  - between the GC and any required authority having jurisdiction (AHJ), and
  - between the GC and subcontractors, chaired by the GC's project manager (PM);
- Safety meetings led by the GC's superintendent or safety inspector;
- Foremen meetings led by the GC's superintendent;
- Subcontractor coordination meetings led by the GC's PM;
- Weekly owner–architect/engineer–contractor (OAC/OEC) meetings led by the GC's PM; and
- Weekly change order, RFI, and submittal coordination meetings.

We will utilize the weekly OAC/OEC meeting as the basis for this discussion on meeting notes; the processes and products are similar for most meetings. This meeting should follow a written agenda. Use of the previous meeting's notes provides an easy and natural path to follow. There are many additional uses for meeting notes, including:

- List of meeting attendees,
- Record of meeting discussions and decisions,
- Record of pending items from previous meetings that were discussed,
- List of action items and due dates,
- Transmission of any handouts distributed during the meeting, and
- Announcement of next meeting.

Customarily, the GC's project manager will chair the weekly OAC/OEC coordination meeting and is responsible for publishing meeting notes. A meeting may be too large for the project manager to both chair and record meeting notes. In such instances, a project engineer may assist with taking the notes. Notes should be prepared and distributed the same day that the meeting is conducted, which serves as timely notification for decisions and open action items.

   Similarly to the daily diary, meeting notes provide a formal historical record and are relied upon heavily in case of a dispute. Meeting notes often serve as supportive data for letters, change orders, and RFIs. They should be filed and stored along with the project files. Meeting note formats may occasionally vary according to specific projects, topics, and attendees. Individual project meeting notes used within a construction firm should follow the same format throughout the firm and often require the following items/processes:

- Heading: Project name, date/time/location of meeting, and identification of the note taker;
- Attendance list and the firm each attendee represents;
- Each new business note item should be numbered with that meeting number as its prefix. Open or unresolved items should be carried forward as old business. The meeting number prefix also indicates the age of open issues. Some firms use the meeting date as the item number;
- Decisions and action items;
- Dates actions are due and individuals responsible for each action;
- Any handouts and logs should be referenced and, if at all possible, attached, regardless of source. This way the project manager can avoid an individual indicating that he or she did not

receive a copy of an item distributed during the meeting. This includes the RFI log, submittal log, ASI or sketch log, and change order proposal (COP) log; and

• The notes should be distributed to all attendees, those absent, and any others who are affected by the notes.

Record documents

There is not just one set of drawings and specifications on any typical project. There likely are many sets, and some of them need to be retained as record drawings. There are several different types of record drawings that the GC will utilize and retain, including:

• The set of drawings estimated and bid from;
• The documents referenced in the original contract;
• The permit set of drawings, as stamped by the municipality—this set is required to be available in the contractor's office for review by the inspector;
• Approved submittal drawings;
• The current set of contract drawings, as incorporated into the contract through change orders; and
• As-built drawings.

As the project progresses, new drawings are customarily issued and should be incorporated into the prime contract through change orders. A *drawing log* should be kept; it should list each drawing, every revision number, and the dates they were issued. This log may be maintained by the design team, the construction team, or both. It should be periodically distributed to make sure that all contracting team members have the current set of drawings and that these drawings have also been incorporated into all subcontracts and purchase orders. The requirements for posting and storing different types of record drawings are usually defined in the front-end of the specifications (CSI Divisions 00 and 01) or in the special conditions to the contract.

As-built drawings are the final set of record drawings and will also be discussed in Chapter 14. As-built drawings are developed by marking up final locations of installed materials on the construction drawings. These important drawings will be used by the owner as a reference for how the project was actually constructed and are valuable for maintenance and future revisions. The superintendent is responsible for maintaining the as-built drawings to ensure that they reflect any changes made during the construction of the project, but this is usually done with help from a PE. The as-built drawings are often "red-lined" on the current contract set. Approved RFIs and any sketches or ASIs will be taped to the back of the preceding drawing as well. If all documents are in electronic format, a project engineer may lead the preparation of the as-built drawings, but they should have input and support from the superintendent and foremen who installed the materials.

Use of technology

Document control is probably the task that has been the most affected by technology changes. Whereas paper documents are often required or recommended for legal reasons, most contractors handle project documents electronically. Documents are either stored on project or corporate databases that act as repositories or on the cloud services provided by many technology vendors. Updating of contractual documents also heavily relies on technology. As an example of how project engineers rely on technology for some of their tasks, the next section discusses the process of electronically recording RFIs on contractual documents.

Electronic posting of RFIs to drawings

Once RFIs have been returned to the GC, this information should be annotated on the drawings so that anyone using them would know about the newly available information. Performing this task electronically provides the advantage of having a single, electronic, set of drawings for anyone to access. However, this also means that all printed sets of drawings will need to be replaced periodically. As shown in Figure 11.4, the RFI process includes five steps. As discussed in Chapter 10, the initial two steps are designed to originate the RFI, which is the duty of the construction team. The third step is designed for the owner and their consultants to respond to the RFI. The last two steps are designed to close the RFI once a response is returned and are the duty of the construction team. The RFI closure process instructs how to post RFIs on contractual documents and when the RFI can be considered closed on the RFI log, as shown in Figure 11.4. This process starts once the project engineer has obtained a response to his or her RFI from the project owner or once their consultants have reviewed the response and found it satisfactory. If the response is not satisfactory, the project engineer needs to evaluate if he or she needs to reformulate the question through a revised RFI or simply resubmit the same RFI to obtain a complete response.

If the response is satisfactory, it is necessary to amend the current set of drawings by incorporating the new information into it for all to use. This is achieved by posting the RFI on the documents. A similar process can be used to post other changes or information, including ASIs. Depending on the level of technology implementation by the construction firm, this task may result in a new set of drawings fully or semi-integrated with the new information. Semi-integration can be achieved by highlighting the portion of the drawings that is affected by the new information and then adding a stamp with the RFI tracking number that points toward the clouded areas. This is shown in Figure 11.5. By reviewing the updated set of drawings, any member of the construction team who is planning work in the clouded area will know that he or she also needs to review the RFI and the response before moving forward with construction activities. The RFI and its response will be stored either in a file cabinet or in an electronic database.

Some contractors prefer full integration, which is achieved by creating a file that incorporates both the RFI and its response and then hyperlinking it to the RFI stamp that is placed in a project server or on the cloud services provided by many technology vendors. Now, by reviewing the updated set of drawings, any member of the construction team that is planning work in the clouded area can click on any RFI stamp and automatically access the RFI and its response. A fully integrated set of drawings allows a reduction in the steps necessary to access information. However, successful implementation also relies on a robust jobsite network that allows all members of the construction team to access information and allows RFI documents to be accessible as needed.

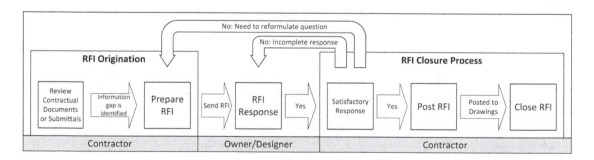

Figure 11.4 RFI Management Process

Project engineering applications

Similarly to what we discussed in chapter 10, management of the submittal process is one of the major areas of involvement for project engineers. The early development of a submittal log, submittal schedule, and incorporation of submittal requirements into each subcontract and purchase order will improve the chances of quality and schedule success for all construction team members. After award, the PE will expedite receipt of submittals, review them to verify that they conform to contract expectations, and forward them on to the design team.

Although authoring of the daily diary should be the responsibility of the superintendent, the project engineer can assist with that process by collecting subcontractor diaries, validating manpower and equipment on site, recording schedule accomplishments, and distributing the completed diary to appropriate parties. Similarly, publication of the weekly OAC/OEC coordination meeting notes is the responsibility of the project manager, but assisting with notetaking during the meeting, documenting attendance, preparing logs to be distributed, attaching those logs to the meeting notes, and distributing meeting notes are all activities for which the PE will support the PM.

There are a variety of record documents on any construction project. One typical role for the PE is to post RFIs, ASIs and sketches, and other design-changing information on to the current record

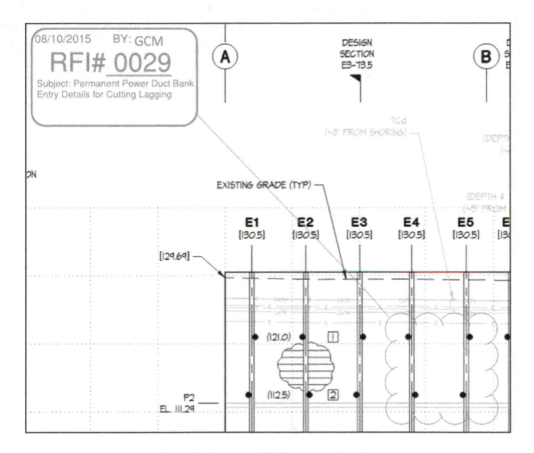

Figure 11.5 Sample RFI Posted on Drawings

document set in the contractor's office. This task is predominantly performed with computer aid, as shown in the previous section.

Summary

There are many different types of construction management documents, in addition to the contract documents that were discussed in Chapter 5 as well as this chapter. The creation, maintenance, and control of construction documentation is the primary focus of project managers, superintendents, and project engineers. Foremen and construction craftsmen build the project with their hands, supported by tools and construction equipment. Project managers, superintendents, and project engineers utilize construction documentation for their daily management tools. Chapters 10 and 11 have discussed a few of the construction documents; many others are referenced throughout this textbook.

Submittals are a continuation of the design process by the construction team, at least through the early to middle phases of the construction control and execution phase. Submittals provide additional detail to the contract drawings and specifications, which provide the designer and owner with assurance that the GC, subcontractors, and suppliers will all be providing and installing the correct materials. Submittals take on a variety of formats, including shop drawings, mock-ups, cut sheets, and physical samples. In addition to validating the design intent, submittals are an early form of quality control and schedule assurance. Project engineers play a major role in RFI and submittal coordination and the corresponding log maintenance.

The superintendent's daily diary is one of the most important and authentic documents on a construction project. It is prepared at the completion of each day by the GC's project superintendent, as well as by all subcontractors' superintendents. It is a contemporaneous record of jobsite progress. Similarly, meeting notes, especially those from the weekly owner–contractor coordination meeting, are an important construction management record authored by the GC's project manager, often with help from one of the project engineers. In the case of a legal dispute, the diary and the meeting notes are among the first and most important documents collected by attorneys. They must both be kept accurate and objective.

There are several different types of record documents, from those the contractor originally prepared in its bid, to the drawings the City issued the permit against, to those drawings referenced in the prime agreement. The most current set of drawings and specifications that the construction team is building are record drawings; these require ongoing markups incorporating actual field conditions and direction from the design team. The final set of drawings, reflecting the as-built conditions, is prepared by the contractors and will be turned over to the project owner for long-term maintenance and future building modifications. The project engineer will play an active role in keeping the current record set of drawings up to date, as well as inputting to the final as-built set of drawings.

Review questions

1. Why are submittals (e.g., shop drawings, product data, or product samples) required from suppliers?
2. Who reviews vendor submittals in the GC's office, and what is he or she looking for?
3. Who (which firm) is responsible for preparing submittals?
4. Who (which firm) is responsible for approving submittals?
5. What is the difference between a contract drawing and a shop drawing?
6. List five items that should require preparation of a submittal on a wooden-frame multi-family construction project, beyond those discussed in this chapter.

7. Why would a submittal be considered part of the quality control process?
8. Why would a submittal be considered part of the design process?
9. When is the submittal log reviewed?
10. Who on the GC's team often prepares or drafts meeting notes?
11. Who on the GC's team is responsible for the accuracy of the weekly owner–designer coordination meeting notes?
12. When should the meeting notes be distributed and why?
13. List at least five important types of information that are recorded on a daily job diary.
14. Who should originate a diary, and why?
15. Why would it be important for a drawing log to be up to date?
16. How are as-built drawings created?

Exercises

1. Prepare a submittal coversheet for the green roof for the Rose mixed-use apartment case study project. Use the project schedule and other documents included on the companion website to assist in identifying approximate dates and lead times.
2. Prepare a submittal coversheet for an item on the I-90 case study project that may not normally require a submittal. Select an item that may involve some uncertainty. Explain why you selected the item.
3. Prepare a set of meeting notes for your classroom discussion as if it were a construction meeting.
4. Let's hope this doesn't happen on your projects, but how might a diary be misused or misrepresented? What would be the ramifications of this?
5. Create a typical job diary for either or both of our case study projects for the following dates. Use the project schedule on the website for reference. Make any reasonable assumptions:
   (a) January 21, 2019 (zero degrees and an ice-storm in Montana, and/or rain for 45 days straight in Oregon),
   (b) August 21, 2018 (a clear day in Oregon, and/or 100 degrees in Montana).

Reference

WSDOT. (2017, August 7). WSDOT Standard Specifications for Road, Bridge, and Municipal Construction. Retrieved September 27, 2017, from www.wsdot.wa.gov/Publications/Manuals/M41-10.htm

12 Cost engineering

Introduction

In previous chapters, we have discussed various project control activities, including safety control (Chapter 7), cost and schedule control (Chapter 8), quality control (Chapter 9), and document control (Chapters 10 and 11). In this chapter, we will integrate all of those controls into jobsite field work planning, including cost control activities beyond those introduced in Chapter 8. Pre-construction planning and scheduling are project management or advanced project engineering functions that are performed with the support of the superintendent. Here we will focus on integrated cost engineering plans and operations accomplished by the project manager, project engineer, and project superintendent. These plans are both fixed and fluid. Contractors have fixed estimates, schedules, and installation teams, but all elements often require adjustment and modification as the project progresses.

In this chapter, we elaborate on the integration of project control activities with planning for field work. Earned value analysis is one method that combines cost and schedule and compares plans to actual performance. Each project must report on all these control operations to upper management, their home office, and sometimes the project owner. Some of those reporting tools include the monthly cost and schedule forecasts, end-of-project as-built estimates, and as-built schedules. Although a lot of the home-office reporting is spearheaded by the project manager, project engineers play a vital information gathering and support role.

Integration of cost, time, quality, and safety goals into construction work

In previous chapters, we have discussed the four important dimensions of construction project success, which must be integrated cohesively on a construction project and act as the pillars to success. Project controls are designed in line with these pillars:

Project Success = f (Cost, Time, Quality, Safety)

This equation was depicted in Figure 2.1. A thorough work plan will address each of these elements individually and then integrate them holistically. Each element has its own goals that the project team must endeavor to achieve, but a truly successful construction project will require all of the elements to successfully work together. To this end, control processes are shaped to support the pillars; if one control process fails, that particular work package or even the entire project might fail.

Cost control starts with a detailed and accurate estimate, which includes and separates subcontractor-performed from general contractor (GC) performed work. The estimate should be assembled and cost-coded by systems, such as doors, frames, and hardware. Time control is managed through schedules. The contract schedule should also be arranged in systems and be detailed enough to allow subcontractor interfaces and the assembly of three-week look-ahead schedules. This approach to formatting the estimate and schedule allows the project engineer and superintendent to efficiently plan detailed construction operations.

Quality and safety control are major focuses of the project superintendent. Many construction firms have corporate quality and safety control officers. Many larger projects, such as our two case studies, have quality and safety control inspectors on the jobsite; sometimes, this is a contract requirement. Smaller companies and smaller projects will rely on the superintendent to supervise these areas along with his or her other responsibilities.

Timely and efficient processing by the project engineer of requests for information (RFIs) and submittals, along with accurate subcontracts and purchase orders is required before any construction can begin. If construction for a work package has begun, but some issue arises that could have been solved earlier, it will likely disrupt construction execution. A workable plan requires integration of all these control functions. For instance, if construction for structural steel has begun, but a dispute with the supplier on a purchase order for the crane rental, a delayed RFI response on weld thickness, or the lack of a jobsite safety plan submission by the erection subcontractor occurs, the project will probably suffer cost and schedule impacts.

In addition to integrating all of the control functions, the superintendent must also assemble several individuals and construction companies into an efficient jobsite team. On large projects, it is common to see hundreds, if not thousands, of craftsmen and up to 100 subcontracting firms and suppliers. This adds to the complexity of the project, because it is not uncommon that 80% of these firms have not worked together on a prior project and they may have different goals in mind for their firms than the general contractor's (GC's) goals. Still, subcontractors are vital members of the project team, and the superintendent must find a way for both them and the GC to be successful in order for the project to succeed. A construction organization chart should be thought of as a complicated spider's web. If any of the main cords of the web are torn, the web will fail. We introduced different individual project roles in Chapter 3. Figure 12.1 shows a sample project organization chart that emphasizes the field side of the project. These firms and individuals must all be successfully coordinated by the GC's project team, especially the general contractor's superintendent.

Some general contractors feel that they have more control of cost, schedule, quality, and safety with their own crews, rather than subcontractor crews. This is especially true on heavy-civil projects, where ownership of the right equipment fleet is sometimes a prerequisite to get the contract. However, the GC will have more flexibility with crew choice on private projects, when procurement includes some negotiation. This is not usually possible on most public projects, when the contract and, usually, the subcontracts are awarded to the lowest bidder. As will be discussed in Chapter 19, several factors go into the selection of subcontractors and suppliers on projects when contracts are negotiated, similarly to the situation of the GC–owner relationship, whereas under a low-bid procurement, subcontracts are awarded almost exclusively to the low-bidding subcontractor. There are advantages and disadvantages of self-performed versus subcontracted work and most experienced construction professionals can effectively argue either side.

Earned value management

Earned value management (EVM) analysis is a technique used by some contractors to determine the estimated value of the work completed to date, also known as *earned value* (EV), and compare it with the actual cost of the work completed. EVM can be applied to analyze the project as a whole or each activity individually.

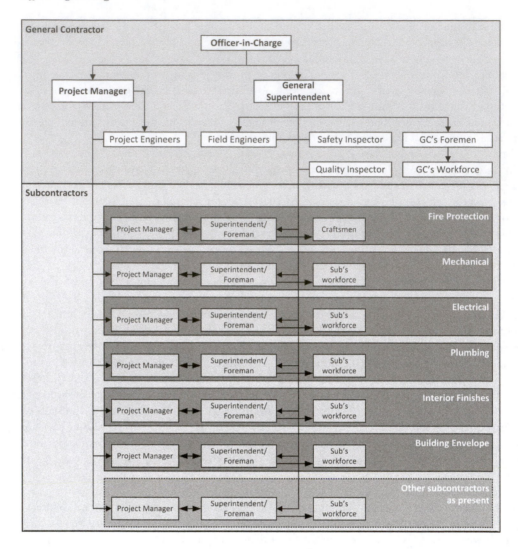

Figure 12.1 Project Organizational Chart

The most effective use of earned value to a general contractor's jobsite project team is to track direct labor, which represents the construction contractor's greatest project risk. Both man-hours and labor costs can be used to perform an EVM analysis. However, man-hours, not labor cost, is best for monitoring direct labor, because counting hours from timecards is easier and does not have to take into account different hourly rates for a wide mix of apprentices, journeymen, and foremen.

EVM is an additional tool to evaluate how the project is doing against the initial plan. For instance, the work-package curve developed in Chapter 8 and shown in Figure 8.2 reflected the man-hours planned by the foreman and the actual man-hours used for the I-90 case study pile-cap installation. Whereas this curve provided useful information at completion, it did not answer some of the crucial questions. During the first week when the actual man-hour curve was below the estimated curve, did this necessarily indicate that the foreman was under budget? Could he or she have been behind schedule? The project team could actually be ahead or behind schedule and over or under budget because the work package compared only estimated to actual hours and did not consider the amount of work accomplished.

Therefore, by adding a third, or earned value curve, EVM combines forecasted effort with actual effort in man–hours and schedule performance. This allows the project team to analyze the cost and schedule performance as a whole. The EV curve plots the quantity of work performed against the hours that were originally estimated for installing that quantity. Therefore, at any given time, the value necessary to plot the EV curve for an activity is equal to the estimated productivity in man–hours per unit multiplied by the number of units in place at the time of reporting. If the cumulative percent completed is known, this curve is determined by plotting the total number of man–hours estimated for the work package multiplied by the cumulative percent completed.

Figures 12.2 and 12.3 include the earned value analyses for the door, frame, and hardware (DFH) installation in the 1st and 2nd floors of the Rose Collective mixed-use apartment case study project. If the field engineers have used field measurements and estimated productivity rates to find that 32% of the work has been installed by the end of day four, then 32% of the estimated hours have been "earned." In other words, 112 of the total 352 hours estimated have been earned, so 112 is plotted on day four. This method of monitoring will provide more accurate feedback to the project team for appropriate cost and schedule reporting and correction if required.

Two different metrics of measurement are now available by using all three curves in concert: schedule and cost variances. By analyzing the values of these variances, EVM allows the project team to answer the following questions:

- Are we right on schedule, ahead of schedule, or behind schedule, and by how many days?
- Are we right on budget, over budget, or under budget, and by how many hours?

Schedule status

The schedule status is evaluated by using the schedule variance in man-hours (MH). This variance is determined by subtracting the estimated effort in MHs from the effort necessary to achieve the earned value.

Schedule Variance (MH) = Earned Value (MH) − Estimated/Scheduled (MH)

Note: As we mentioned before, in an effort to use gender-neutral language to reflect an increasingly diverse workforce, some companies are replacing the term man-hours with person-hours (PH). This terminology is less common and not widespread, so we decided to follow the traditional terminology in the rest of this book.

If this variance returns a negative value, it means the project or activity is behind schedule. If it returns a positive value, it means the project or activity is ahead of schedule.

For instance, DFH installation on the 1st floor of the mixed-use development (MXD) project is returning a negative schedule variance in MHs on day six, which means the activity is behind schedule. To determine how many days behind schedule, we will need to compute the schedule variance in time units, which is calculated by measuring the horizontal distance between the earned value and estimated curves, as shown in Figure 12.2. This graphical analysis is suggesting that this activity is roughly one day behind schedule.

Conversely, DFH installation on the 2nd floor of the MXD project is returning a positive schedule variance in MHs on day six, which means the activity is ahead of schedule. To determine

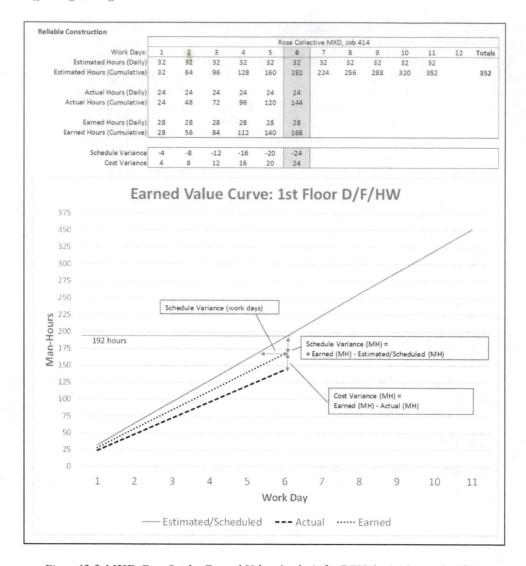

Reliable Construction

		Rose Collective MXD, Job 414											
Work Days:	1	2	3	4	5	6	7	8	9	10	11	12	Totals
Estimated Hours (Daily)	32	32	32	32	32	32	32	32	32	32	32		
Estimated Hours (Cumulative)	32	64	96	128	160	192	224	256	288	320	352		352
Actual Hours (Daily)	24	24	24	24	24	24							
Actual Hours (Cumulative)	24	48	72	96	120	144							
Earned Hours (Daily)	28	28	28	28	28	28							
Earned Hours (Cumulative)	28	56	84	112	140	168							
Schedule Variance	-4	-8	-12	-16	-20	-24							
Cost Variance	4	8	12	16	20	24							

Figure 12.2 MXD Case Study: Earned Value Analysis for DFH Activities on 1st Floor

how many days the work is ahead of schedule, we will again need to compute the schedule variance in time units, which is computed by measuring the horizontal distance between the earned value and estimated curves, as shown in Figure 12.3. This graphical analysis is suggesting that this activity is roughly one day ahead of schedule.

Cost status

The cost status is evaluated by using the cost variance. This variance is determined by subtracting the actual effort in man-hours from the effort necessary to achieve the earned value.

Cost Variance = Earned Value − Actual Cost

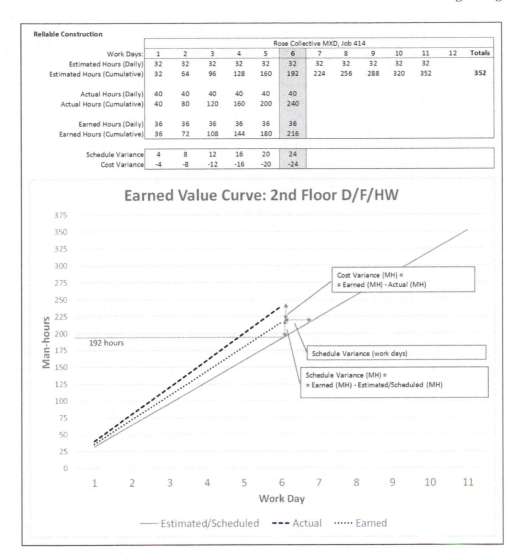

Figure 12.3 MXD Case Study: Earned Value Analysis for DFH Activities on 2nd Floor

If this variance returns a negative value, it means the project or activity is costing more than estimated. If it returns a positive value, it means the project or activity is costing less than estimated.

For instance, DFH installation on the 1st floor of the apartment project is returning a positive cost variance on day six, which means the activity is costing less, because it has taken 24 fewer man-hours than expected to achieve the earned value, as shown in Figure 12.2. Conversely, DFH installation on the 2nd floor of the apartment project is returning a negative cost variance on day six, which means the work activity is costing more, because it has taken 24 more man-hours than expected to achieve the earned value, as shown in Figure 12.3.

Home-office reporting

The degree of autonomy any jobsite project team has from the home office varies depending upon the culture of the construction company, the complexity of the project, contract terms, and the individual team members involved. Nevertheless, there will always be input and guidance from the home office to the jobsite and reporting accountability from the jobsite back to the home office. Many construction firms require the project manager (PM) to prepare a monthly report to management. This report summarizes results against the four dimensions of project success, as well as other critical project issues, such as subcontractor performance and project owner and/or designer progress on their own duties, including processing of RFIs and change orders. The superintendent will contribute significantly to the quality and safety portions of that report with assistance from the project engineers. Three methods of home-office reporting include monthly cost and fee forecasts, end-of-project as-built estimates, and as-built schedules.

Forecasting

The project manager is responsible for developing a monthly cost forecast for the project that will be shared with the officer-in-charge and the executive management. In preparing this report, the project manager often relies on the assistance of the project engineer, field foremen, and superintendents. The contractor's bonding and banking partners may also have an interest in the monthly forecast. The project owner may be sent a copy of the monthly forecast in the case of a cost plus contract that relies on open-book accounting. This forecast includes line items for all areas of the estimate, cost to date, and estimated cost to complete. Each of the major areas of work receives a separate forecast page, and each of those is broken down for all categories of work. The major categories of the estimate include:

- Direct labor,
- Direct material,
- Subcontractors and major material suppliers,
- Jobsite administration or general conditions, and
- Percentage markups, including fee, excise tax, contingency, and insurance.

Table 12.1 illustrates only the summary forecast page for the Rose Collective case study project. The complete forecast is often as long as the original detailed estimate, possibly 20 pages and hundreds of line items. It is a best practice to include a narrative with the monthly forecast explaining the significant differences from the previous month's forecast, as well as a work plan for continuing or improving performance throughout the remainder of the project. The management team cannot afford to wait until the project is complete to measure and report the overall project cost. It is not only too late to take corrective action, but it is also too late to accurately determine why the team deviated from the plan.

There can be several other management reports, either generated by the project manager or the accounting department, according to the construction firm's practices and the requirements of any specific owner or project. Popular reports include the weekly labor report, monthly job cost history report, equipment log, and accounts payable report. All these reports are computer-generated and are accurate to the degree that the information regarding actual costs was accurately input. These reports can occur weekly or monthly, but either way, home-office-generated cost control reports are likely to be too late for implementing any corrective action in the field. However, they are needed in the home office to evaluate the company exposure and success across all of its projects.

Table 12.1 MXD Case Study: Monthly Forecast Summary

Reliable Construction Company

MONTHLY FORECAST SUMMARY PAGE

Project: Rose MXD   Job No.: 414   PM: Jason Campbell   Date: 1/1/2019

CSI Division	Description	Estimate Totals	Change Orders	Current Contract	Cost To-Date	Cost To-Go	Forecast Cost	Variable +/-
1	General Conditions	$1,573,437		$1,573,437	$629,375	$944,062	$1,573,437	$0
2	Sitework	$1,744,453		$1,744,453	$1,525,000	$275,000	$1,800,000	–$55,547
3	Concrete	$2,097,716	$22,250	$2,119,966	$1,650,250	$520,000	$2,170,250	–$50,284
4	Masonry	$0		$0	$0	$0	$0	$0
5	Structural Steel	$1,001,353		$1,001,353	$150,000	$875,555	$1,025,555	–$24,202
6	Wood	$2,355,625	$75,003	$2,430,628	$201,003	$2,175,000	$2,376,003	$54,625
7	Thermal and Moisture	$1,494,665		$1,494,665	$175,000	$1,319,665	$1,494,665	$0
8	Doors and Windows	$1,194,164		$1,194,164	$222,300	$950,125	$1,172,425	$21,739
9	Finishes	$1,782,986		$1,782,986	$0	$1,505,000	$1,505,000	$277,986
10	Specialties	$115,183		$115,183	$0	$115,183	$115,183	$0
11	Equipment	$462,830	$100,000	$562,830	$75,000	$487,830	$562,830	$0
12	Furnishings	$31,305		$31,305	$0	$31,305	$31,305	$0
13	Special Construction	$0		$0	$0	$0	$0	$0
14	Elevators	$472,000		$472,000	$100,000	$372,000	$472,000	$0
15	Mechanical	$1,686,879		$1,686,879	$375,000	$1,301,002	$1,676,002	$10,877
16	Electrical	$1,264,850		$1,264,850	$85,000	$1,179,850	$1,264,850	$0
	Total Cost:	$17,277,446	$197,253	$17,474,699	$5,187,928	$12,051,577	$17,239,505	**$235,194**
	Fee @ 5%	$863,872	$9,863	$873,735				**Saving**
	Other Markups	$557,469	$7,890	$565,359				
	Total Contract:	$18,698,787	$207,116	**$18,905,903**				

Forecast Loss: $0
Forecast Savings: $235,194

GMP Savings Split: 80% Client, 20% GC = $47,039

Current Contract Fee: $873,735

Forecast Final Fee: **$920,774**

As-built estimates and schedules

When a project is close to completion, project managers and project engineers are often too busy and excited about starting the next project to develop as-built estimates for a completed project. Accurate as-built estimates, like as-built drawings and as-built schedules, are important historical reporting tools for a construction firm. Revision of the estimate with actual cost data assists in developing better future estimates. This is particularly true if actual unit price data, which require input of actual quantities as well as actual cost, are determined. These data can help with developing the firm's database and provide project managers and estimators with accurate cost factors for future use. Updating of the schedule and estimate and submission of relevant inputs to the home office, including as-built estimates and as-built schedules, will also be discussed in Chapters 14 and 16.

Just as an as-built estimate is important for estimating future work, an as-built schedule is helpful in scheduling future projects. Similarly to estimating, the greatest risk in scheduling is the determination of direct craft workforce productivity. The project manager or home-office scheduler should develop a personal set of productivity factors based on actual prior experience and historical data. These factors will help future jobsite project teams establish realistic activity durations when scheduling their next projects.

Use of technology

Cost engineering tasks rely highly on technology. Database applications are used to support cost engineering functions for a variety of different purposes. There are several software systems available that allow a contractor to maintain a customized database and prepare reports for multiple uses. Moreover, data visualization tools are being used to automatically develop weekly reports and monthly progress reports, as shown in Figures 12.4 and 12.5. Figure 12.5 shows the evolution

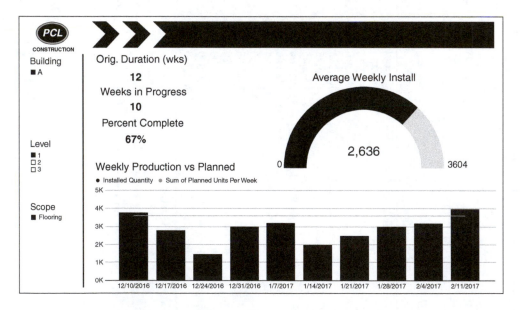

Figure 12.4 Weekly Report using Microsoft Power BI
(Credits: Matt Glassman, PCL Construction)

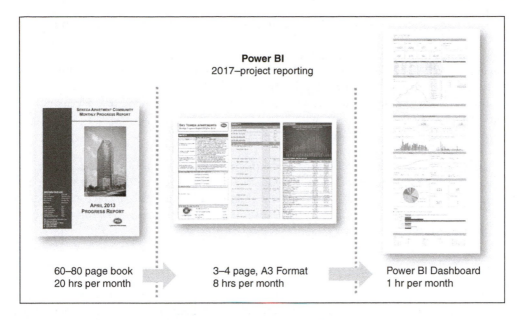

Power BI
2017–project reporting

60–80 page book
20 hrs per month

3–4 page, A3 Format
8 hrs per month

Power BI Dashboard
1 hr per month

Figure 12.5 Transition from Written Monthly Progress Reports to Web-based Automatic
Reporting with Microsoft Power BI
(Credits: Matt Glassman, PCL Construction)

of reporting made by a major general contractor and how the use of databases and data visualization tools, such as Microsoft Power BI, have allowed this company to reduce the effort required to produce monthly reports.

Project engineering applications

Traditionally, college graduates in civil, architectural, or construction engineering or construction management who decided to work for a construction firm were placed on a typical career path. This led to advancement into a project management position after several years of experience as a project engineer in the field and in the jobsite office. However, many contractors are finding it useful to identify those employees who are more interested in field activities and to help them mature into a career path that leads to them becoming construction superintendents. Whether one's goal is to become a project manager or a superintendent, previous experience in working in the field as a carpenter, laborer, or ironworker can be preparatory for success in this career path. Comprehension of the planning necessary to construct an assembly and what materials, tools, and equipment are necessary is an extremely valuable asset. Project engineers, whether situated in the office or in the field, will be more valuable to their firms and will progress faster in their careers if they have this understanding and a background in construction assemblies. Project reporting is a major duty for site managers, such as project managers and superintendents.

The jobsite project engineer will support the superintendent with work planning activities and reporting in a variety of fashions, including:

- Making sure purchase orders and subcontracts are written in a timely and complete manner,
- Processing RFIs and submittals before construction of a system begins,
- Expediting material deliveries so that materials are ready for foremen when needed,

- Assisting the project manager with transitioning from foreman work packages to earned value studies and analysis,
- Forecasting direct work quantities and man-hours to-go, and/or
- Preparation of as-built estimates, which are not simply reporting actual costs but require the input of actual installed quantities to determine as-built unit pricing. The same holds true for as-built schedule development. It is a function of project engineering to collect final installation quantities and compare them with actual material costs and actual man-hours, to prepare as-built estimates and as-built schedules.

The project engineer will also support the project manager with assembly of cost, schedule, quality, and safety control data to be reported to the home office. This may be done in a formal monthly report, potentially accompanied by a presentation and site visit from the contractor's officers. More advanced project engineering support includes assistance with change orders and pay requests, as will be introduced in later chapters.

Summary

As well thought out as any pre-construction plan may be, it typically does not go into enough detail for the jobsite management team, particularly the superintendent and project engineer, to manage the construction work. Assembly of work activities into standalone work packages, also known as assemblies or systems, is more efficient for both planning and reporting progress on construction work than planning with the CSI categories. Many elements go into the assembly of a complete work package, including material, tools, equipment, cost estimate, schedule, and subcontractor interfaces. In the field, the project superintendent must integrate the original estimate and contract schedule with the project-specific quality and safety control plans, along with documentation backup. This is all accomplished with the assistance of the project engineer. A more advanced technique, utilized for cost and schedule control, is earned value management analysis. This compares the scheduled and estimated value of work completed with the actual duration and cost incurred.

The project manager receives input on all of the project controls from his or her team members and keeps the contractor's home office informed. Often separate reports, be they weekly or twice monthly, will be prepared for each of the controls. Sometimes a monthly status report is prepared that covers these issues, as well as client and designer relations. The home office cannot wait until the project is complete to find out if the bid fee has eroded, the schedule is delayed, someone was hurt, or the client's quality expectations were unfulfilled. An efficient control system allows for correction and modification if timely communication and document tools are in place.

One of the most important cost reporting tools generated by the PM is the monthly cost forecast. It is essentially a re-estimate of the entire project on an activity-by-activity basis, incorporating costs incurred to-date, schedule status, and actual productivity rates and forecasting the costs and time remaining to complete the project. As-built estimates are developed to create unit price data that can be used on future estimates. As-built schedules are prepared to allow schedulers to develop historical productivity factors for use on future projects.

Review questions

1. What is the difference between pre-construction planning and jobsite construction work planning?
2. What are as-built estimates used for?
3. What are as-built schedules used for?

4. What is an earned value analysis used for?
5. Of the four control pillars, which was most likely "originated" before the project started by the project manager, project engineer, or superintendent?
6. Of the four control pillars, which will most likely be "controlled" or managed after construction begins by the project manager, project engineer, or superintendent?
7. How frequently should a project manager develop a cost and fee forecast?
8. What percentage of the work was scheduled to be complete by day four in Figure 12.3, and how much had been accomplished?

Exercises

1. Prepare a cost forecast for the work activities from Chapter 8, Exercise 9, based upon the work packages already developed. Forecast the remaining hours needed and calculate the total under- or overrun this system will achieve. Convert the hours to dollars by using the wage rates from Chapter 17 or other local current rates. List out the possible reasons why the under- or overruns could be occurring and what corrective actions should be taken. If all continues on this same trend (proceeds at the same rate as it has been), calculate what the historical as-built unit man-hours will be.
2. Prepare additional earned value curves by starting with the work package example presented in Chapter 8, Figure 8.2. Draw a curve where the work is over budget and behind schedule, and another where the team is on budget and ahead of schedule. Prepare a narrative explaining why these situations might be occurring.
3. Referring to the earned value curves in Figures 12.2 and 12.3, if each of these work packages proceeds on the same trend, when will they be complete and will the final hours be over or under budget?
4. Of the four control pillars, which do you feel is the most important and/or least important to these stakeholders? Note that there are not any exact answers here.
   a. Project owner
   b. Lead architect or engineer
   c. General contractor
   d. Major subcontractor
   e. Authority having jurisdiction

13 Sustainable built environment

Introduction

This book focuses on the built environment, so we will be discussing sustainability as it relates to the design and construction industries and their projects. Although many of the sustainable aspects of a facility or infrastructure project are a result of how design decisions affect operations, other aspects relate to the manner in which the construction is performed. Achievement of sustainability in the built environment means the planning and execution of a construction project in such a manner as to produce a facility or infrastructure that fits in its social, economic, and natural environments, as shown in Figure 2.3F. At the same time, efforts to minimize, nullify, or better yet reverse any adverse impact of the construction process on the natural environment, economy, and society will contribute to sustainability.

As shown in Figure 13.1, sustainability is often represented by a three-legged stool, often known as the *triple bottom line*, which is based on society, economy, and the natural environment. These dimensions overlap and sometimes have mismatching objectives. For instance, most people in the 1980s and 1990s would have said they wanted less pollution and healthier buildings, but if business owners could not make a profit, they would not pay the extra cost to use sustainable products in their buildings and operations. Once it was discovered that healthy buildings improve workforce productivity, business owners became motivated to produce better buildings because of the economic return generated by this productivity improvement. All parties involved in built environment industries, including architects, engineers, building officials, owners, contractors, and suppliers are now more inclined to work together to achieve an optimal triple bottom line for their projects. However, it is crucial for a project owner to state their sustainability goals among the project objectives and write contractual documents that clearly convey them. This allows designers and contractors to develop the project along these guidelines.

However, evaluation of the sustainability of a project is a difficult task. Economic and environmental attainments are usually quantifiable, but social attainments are difficult to directly link to a built environment project and often cannot be quantified. Moreover, construction firms can only minimally contribute to the broader economic and societal attainments. Examples of this involvement include construction firms' efforts to achieve economic soundness by staying on budget, reduce societal impacts by creating a safe and healthy work environment, and limit impacts on businesses and residents affected by the project construction. These attainments are discussed in several chapters of this book outside the context of sustainability. On the other hand, environmental attainments will be the major focus of the rest of this chapter.

The built environment is a major contributor to domestic greenhouse gas (GHG) emissions. Whereas statistics specific to infrastructures are not available, the most recent statistics from the U.S. Environmental Protection Agency (EPA) show that buildings contribute 13% of the domestic GHG emissions. This statistic excludes the emissions from electricity use in buildings, which are

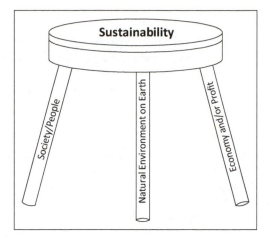

Figure 13.1 Triple Bottom Line

compiled under a different category, Electricity and Heat Production. That category also includes emissions from electricity use in industry and cumulatively contributes 29% of the domestic GHG emissions (EPA 2017). Still, it is estimated that, in 2016, the commercial and residential sectors, respectively, purchased 36.6% and 37.9% of the electricity (EIA 2017). Therefore, we can esti- mate that the emissions from electricity use in buildings are around 21% of the domestic GHG emissions, which brings the total emissions from buildings up to 33%. Even if this value excludes infrastructures, it can be concluded that the built environment is a major contributor to domestic GHG emissions.

Environmental attainment of a built environment project can be measured in terms of its actual performance by evaluating the GHG emissions from development and operation of the facility or infrastructure. However, this approach would not allow project parties to control the outcome and would miss other impacts on the natural, social, and economic environments. Environmental sus- tainability in the built environment can go beyond GHG emissions by considering other aspects, such as:

- Reducing *energy consumption* and reliance on fossil fuels,
- Reducing the need for *transportation* of materials,
- Reducing *water use* and protecting natural resources, and
- Minimizing *air pollution* and protecting air quality.

Various systems have been developed to rate projects based on their expected attainment, so that project parties can receive guidance in their decision making during the project phases. In the remainder of this chapter, the concept of sustainability will be defined and contextualized within the built environment. The role and responsibility of design and construction professionals will then be discussed through examples. Subsequently, an overview of various systems for evaluating the sustainability of materials, products, and projects is included. A discussion of how project engineering tasks relate to sustainability concludes the chapter.

Defining sustainability

Before we focus on built environment projects, it is paramount to introduce the concept of sustainability at large. The following expression provides a good approach to introducing this concept:

> *We do not inherit the earth from our ancestors; we borrow it from our children.*

(LEED 2009)

The term *sustainability* pertains to many facets of life and is not new to humanity, but its current meaning is grounded in a 1987 report of the United Nations (UN) World Commission on Environment and Development. At the time, the commission was chaired by G. H. Brundtland and included, as U.S. representative, William D. Ruckelshaus, who had formerly worked under the Nixon and Reagan administrations. This report, titled "Our Common Future," but frequently referred to as the Brundtland Report, introduced the concept of *sustainable development*, which provided the foundation for what is now referred to as *sustainability*. Some excerpts from this report are included and commented on in this section to provide a grounding for policies and practices on sustainability in the built environment that will be discussed in the rest of this chapter.

The Brundtland Report introduces sustainable development as an objective for human development in the 21st century:

> *Humanity has the ability to make development sustainable to ensure that it meets the needs of the present without compromising the ability of future generations to meet their own needs.*

(UN 1987; sect. 27)

Sustainable development is proposed to create a worldwide integrated approach to deal with environmental, economic, and social change in the world:

> *The concept of sustainable development provides a framework for the integration of environment policies and development strategies … The pursuit of sustainable development requires changes in the domestic and international policies of every nation.*

(UN 1987; sect. 48)

As defined, sustainable development does not mean stopping economic growth; instead, it means seeking a new type of growth that would allow meeting the needs and aspirations of future generations while addressing poverty and underdevelopment:

> *Sustainable development seeks to meet the needs and aspirations of the present without compromising the ability to meet those of the future. Far from requiring the cessation of economic growth, it recognizes that the problems of poverty and underdevelopment cannot be solved unless we have a new era of growth in which developing countries play a large role and reap large benefits.*

(UN 1987; sect. 49)

This means that a balanced approach between economic growth and its environmental impacts should be pursued through policies that seek growth while protecting the natural environment. This can be done by proactively focusing on sources of environmental problems, rather than symptoms:

> *Economic growth always brings risk of environmental damage, as it puts increased pressure on environmental resources. But policy makers guided by the concept of sustainable development will necessarily*

work to assure that growing economies remain firmly attached to their ecological roots and that these roots are protected and nurtured so that they may support growth over the long term. Environmental protection is thus inherent in the concept of sustainable development, as is a focus on the sources of environmental problems rather than the symptoms.

(UN 1987; sect. 50)

World diversity is recognized by the report by giving each country the freedom to pursue sustainability through its own policies while accepting that sustainable development should be a globally shared objective; still, the report recognizes that development is aimed toward globalization, and therefore, international relations will be crucial to achieve sustainable development:

No single blueprint of sustainability will be found, as economic and social systems and ecological conditions differ widely among countries. Each nation will have to work out its own concrete policy implications. Yet irrespective of these differences, sustainable development should be seen as a global objective. ... No country can develop in isolation from others. Hence the pursuit of sustainable development requires a new orientation in international relations.

(UN 1987; sect. 51–52)

Sustainable built environment

What makes a building or infrastructure sustainable? There are many steps that owners, designers, and builders can take to build sustainably, starting with choosing the right site. A building site that is within city limits is often more sustainable than development in a rural area on undeveloped land (for example, farm or forest land), because it would have a reduced impact on the natural environment; it already has access to power and water and has an infrastructure nearby. The size of the building's footprint on the site is also a consideration. If the building's footprint is smaller and allows for landscaped areas, less water will run off the site, which will mitigate erosion and/or the spread of pollutants. The use of renewable building materials is another important step toward sustainability. *Renewable materials* are those that can be reproduced at a rate that meets or exceeds the rate of human consumption; renewable materials should not be considered as able to meet infinite demand.

There are many options available to architects and builders so that they can tailor a sustainable building to fit their clients' needs. Some popular choices include:

- Specifying heating, ventilating, and air conditioning (HVAC) systems that do not emit hydrofluorocarbon (HFC) and hydrochlorofluorocarbon (HCFC) gases,
- Installing low-flow faucets and toilets,
- Using light-emitting diode (LED) or compact fluorescent lamp (CFL) light fixtures that are on system timers, and
- Setting up a recycle program, both during construction and after close-out.

Once a sustainable structure is built and the owner or tenants move in, it then becomes extremely important to operate and maintain the building sustainably. The owner has invested in high-quality systems, and if they are not programmed correctly or working efficiently, the expected sustainable attainment will not be achieved. An example would be if the HVAC unit was not balanced or was not properly programmed to heat or cool the building. Most contractors and subcontractors are now offering owners maintenance packages or classes for facility managers. Several rating and certification programs that are discussed later in this chapter require the building's owner to subject their projects to checkups every one to five years to maintain their sustainability rating.

Roles and responsibilities of project parties

The design, construction, and operation of projects have environmental, economic, and social impacts. As previously stated, sustainable construction requires evaluation of these potential impacts and selection of strategies that emphasize the positive impacts while minimizing the negative impacts. The roles of designers, such as architects, engineers, contractors, and suppliers, with respect to designing and building sustainable projects often appear intermixed, especially by owners desiring some type of sustainability certification.

In the building sector, even when the intention of all parties is to produce a sustainable or "green" building for the client, the roles are actually quite distinct. Architects and engineers specify building materials and produce drawings. They interpret building codes and the owner's program and desires, and they represent all that information in a set of contract documents, which were introduced in Chapter 5. Some sustainable efforts by the *design team* might include:

- Identifying a building orientation (i.e., location) on the site that would minimize energy consumption,
- Enhancing local energy generation, such as using solar photovoltaic panels,
- Selecting energy-efficient HVAC and electrical systems,
- Specifying building products and components, such as operable and properly placed windows, to maximize natural lighting and ventilation, local building products requiring less driving and fuel consumption for deliveries, and materials that use less energy in manufacture and in end use, and
- Incorporating recycled materials into the specifications.

The overall sustainability of a project is greatly influenced by the design, but the builder can influence the environmental and social impact of the construction operations. In most cases, contractors do not "choose" materials or where materials are installed. The *construction team's* responsibilities in producing a sustainable building include:

- Demolishing or recycling materials in an efficient manner,
- Providing a bus for craft workers or vouchers for the public bus system,
- Reusing materials and minimizing construction waste,
- Using recycle bins and separating materials on the jobsite,
- Minimizing noise, light, and air pollution during construction,
- Protecting and restoring the natural environment,
- Eliminating storm water runoff and soil erosion,
- Selecting construction materials with high recycled content,
- Providing the exact products specified by the design team, and
- Tracking all paperwork to support the materials installed and processes used.

When the green building movement appeared in the United States in the 1990s, it was often seen as expensive and cumbersome by design and construction organizations. Today, it just makes sense to use products manufactured with renewable resources. The quality, availability, and options among these products have improved exponentially. In the lumber industry, many innovations have occurred, from basic forest management to the lumber manufacturing processes. For example, in the Pacific Northwest, the spotted owl had a significant impact on the farming of old growth forests, especially on Washington State's Olympic Peninsula in the 1980s. The outfall of not being able to use these trees caused the wood products industry to become more creative with using other readily available and renewable resources to not only replicate but improve products. Examples include "engineered" lumber products, such as:

- Trus Joist International® (TJI) floor joists,
- Oriented strand board (OSB) plywood,
- Dove-tailed studs, and
- Glue lam beams (GLB), parallel strand lumber (PSL), laminated strand lumber (LSL), and laminated veneer lumber (LVL).

Sustainability rating and certification systems

At the project level, numerous organizations have developed systems to rate and certify the sustainability level of buildings and infrastructures. Some of these systems are discussed in this section, which is organized around project types.

Commercial and institutional buildings

Once laws and building codes are in place and the triple bottom line is penciled out, owners, architects, and contractors begin looking for ways to measure and verify sustainable construction practices. Many organizations have been created to help in the effort of measuring and verifying sustainable practices. For the commercial built environment, we will elaborate on the Leadership in Energy and Environmental Design (LEED) program, the Living Building Challenge, and Green Globes, which are common measurement methods. We will also discuss alternative systems for homes and infrastructures.

The *Leadership in Energy and Environmental Design* rating and certification system was created by the United States Green Building Council (USGBC), a non-profit organization that was founded in 1993. The demand for commercial buildings to be certified as sustainable was so great that the USGBC (the parent organization) created the Green Building Certification Institute (GBCI). The GBCI is a third-party organization that provides independent review of two programs:

- The certification of green building projects, according to the LEED rating and certification system, and
- The training of professionals in the industry toward achievement of LEED credentials.

LEED was developed to help building owners and operators become more environmentally responsible and use resources efficiently. Certification of a building starts with the project owner's decision to pursue LEED certification early in the design process. As part of the registration process, the owner establishes goals for the project in each of nine categories. Buildings are *certified* via a scorecard and people are *accredited* through classes and a final exam. Several scorecards were created for specific categories of building projects: New Construction, Core & Shell, Schools, Retail, Hospitality, Data Centers, Warehouses & Distribution Centers, and Healthcare. Each scorecard provides a framework for scoring projects on nine categories. The points available for each category vary from one type of project to another, as shown in Table 13.1. Once all the scorecards and paperwork are turned in by the contractor, the GBCI reviews them.

These categories result in a possible total of 110 points. Based on their scoring, buildings can achieve four different levels of certification: Certified (40–49 points), Silver (50–59 points), Gold (60–79 points), and Platinum (80–110 points). With the current technology and practices, achieving a Certified or Silver rating is easy to achieve for an experienced designer and contractor

Table 13.1 LEED Point Breakdown by Project Type

	New Construction	Core & Shell	Schools	Retail	Hospitality	Data Centers	Warehouses & Distribution Centers	Healthcare
Integrative Process	1	1	1	1	1	1	1	1
Location and Transportation	16	20	15	16	16	16	16	9
Sustainable Sites	10	11	12	10	10	10	10	9
Water Efficiency	11	11	12	12	11	11	11	11
Energy and Atmosphere	33	33	31	33	33	33	33	35
Materials and Resources	13	14	13	13	13	13	13	19
Indoor Environmental Quality	16	10	16	15	16	16	16	16
Innovation	6	6	6	6	6	6	6	6
Regional Priority	4	4	4	4	4	4	4	4
TOTAL POINTS AVAILABLE	**110**	**110**	**110**	**110**	**110**	**110**	**110**	**110**

team, whereas Platinum is extremely difficult. Once a building attains a LEED certification, the owner is provided with a plaque that can be placed in the building to show the level of certification, as shown in Figure 13.2. Professionals can also obtain LEED credentials through classes and examinations. There are three levels of credentials on five different focuses, as shown in Table 13.2.

The *Living Building Challenge* (LBC) is a newer rating and certification system that offers the most strenuous benchmark for sustainability: creating a zero carbon footprint on our environment. It utilizes most of the LEED requirements but raises the benchmark even higher. Whereas LEED is focused on minimizing negative impact, LBC rewards buildings that are able to have positive impact. These structures are so efficient that they use only power generated on site and

Figure 13.2 LEED Gold Plaque

Table 13.2 LEED Credential Levels

Level	Credential	Specialty	Eligibility	How to Get it
1	**Green Associate (LEED GA)**		No eligibility requirements	Pass LEED GA exam
2	**Accredited Professional (LEED AP)**	*Building Design + Construction (BD+C) Operations + Maintenance (O+C)*	LEED project experience strongly recommended	No previous LEED GA credential: (1) Pass combined GA + AP exam; (2) Pass GA exam then pass AP exam.
		Interior Design + Construction (ID+C) Neighborhood Development (ND) Homes (HOMES)		Previous LEED GA credential: pass only LEED AP exam
3	**Fellow**		LEED AP who have acquired significant technical knowledge and skills	Voted in by peers

captured rainwater runoff to maintain building operations. All building products must be sustainable and the completed building will have a "zero carbon footprint." A zero carbon footprint for buildings means zero net energy consumption or zero net carbon emissions on an annual basis. These buildings are considered as an important strategy to achieve energy conservation and reduce greenhouse gas emissions. The *Bullitt Center* in Seattle, WA, was the first commercial building to meet the Living Building Challenge. The citizens of Seattle, especially those active in the built environment, were very excited about this achievement by the Bullitt Center; it clearly put Seattle and the Pacific Northwest at the forefront of implementing sustainability in the built environment. The building is six-stories tall and has a total of 50,000 square feet (SF). Solar panels generate all the electricity for the building, and rain supplies the water. There is a water treatment plant on site that treats fresh water and wastewater.

Another rating and certification system, known as *Green Globes*, is distributed and run by the Green Building Institute (GBI). Green Globes was originally created in Canada and is now available in the U.S. The Green Globes rating system includes a web-based self-assessment tool, a rating system for certification of a building, and a guide for enhancing the sustainability of a project. The website is very interactive and allows easy communication during the construction process. Green Globes has very similar sustainability standards to LEED. Its requirements are very straightforward and easy to follow. Its ratings are based on: project management; site; energy; water; resources, building materials, and solid waste; emissions and effluents; and indoor environment. Based on verification by a third party, certification can be granted as one to four globes, with four globes being the highest. Certification levels are based on the percentage of applicable points achieved by the project. Green Globes is slightly more expensive to certify than LEED.

Residential buildings

The residential built environment industry also has several means of measuring sustainable building practices. The *National Association of Home Builders* (NAHB) is a trade association established in 1942 that promotes housing policies. The NAHB analyzes policy issues, works toward finding ways to improve the housing finance system, and forecasts economic and consumer trends.

The NAHB created the National Green Home Building Standard (NGHBS) for residential buildings in 2007. This rating system is for new and remodeled single- and multi-family homes and residential subdivisions. The NGHBS rating system focus is on:

- Energy efficiency,
- Water conservation,
- Resource conservation,
- Indoor environmental air quality,
- Site design, and
- Home owner education.

The *Home Energy Rating System* (HERS) was developed by the Residential Energy Services Network (RESNET) in 2006. The HERS index is another official verification system for energy performance in residential construction and is recognized by the Department of Energy, Department of Housing and Urban Development (HUD), and the EPA as an industry standard. A RESNET inspector tests and evaluates new and existing homes for energy efficiency. Based on the scorecard results, a HERS Index score is determined, as shown in Figure 13.3. The lower the index number, the more energy-efficient is the home.

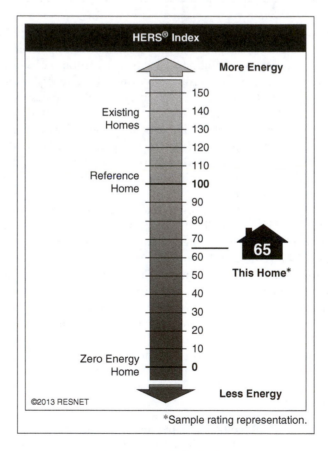

Figure 13.3 HERS Index
(Image courtesy: RESNET)

Infrastructure projects

The *Greenroads Rating System®* was created by researchers at the University of Washington and adapts the certification structure of the LEED system to transportation projects. The system is currently managed by an independent entity, Greenroads International. This entity provides opportunities for sustainability education in the transportation sector and manages the certification process for transportation projects in the U.S. and internationally. The latest version of this system allows projects to earn up to 61 credits:

- 12 mandatory credits (i.e., "Project Requirements"),
- 45 voluntary credits arranged in five Core Categories: (a) Environment & Water, (b) Construction Activities, (c) Materials & Design, (d) Utilities & Controls, and (e) Access & Livability, and
- 4 extra credits that are available in a category called Creativity & Effort.

Projects are assigned a cumulative Greenroads score depending on the points earned. Four certifications levels are currently offered:

- Bronze (40 points minimum),
- Silver (50 points minimum),
- Gold (60 points minimum), and
- Evergreen (80 points minimum).

Another rating and certification system, known as *Envision*, is distributed and run by the Institute for Sustainable Infrastructure (ISI). This system evaluates projects against 60 sustainability criteria to evaluate their environmental, social, and economic impacts. These criteria are organized into five categories:

- Quality of Life (13 credits),
- Leadership (ten credits),
- Resource Allocation (14 credits),
- Natural World (15 credits), and
- Climate and Risk (eight credits).

Also, this system offers four certifications levels, which replicate those in the LEED system but with differing percentages:

- Bronze: To achieve at least 20% of points available,
- Silver: To achieve at least 30% of points available,
- Gold: To achieve at least 40% of points available, and
- Platinum: To achieve at least 50% of points available.

Green material and product certifications

At the material and product level, there are many different certifications that help consumers verify sustainable products that are available for sale, manufacture, and installation in the built environment. Some of these certifications label products based on their performance against a single attribute (e.g., energy consumption), whereas others provide an assessment against multiple attributes. The following are a few of the more well-known product certifications.

The Environmental Protection Agency established *Energy Star* as a voluntary program to identify and promote the use of energy-efficient appliances, heating, and cooling equipment, lighting, home electronics, commercial roofing, and office equipment. The EPA's goal is to reduce energy consumption and pollution. The EPA created a voluntary labeling system for products that meet the highest level of energy efficiency. Several utility companies offer rebates to owners who install energy-efficient materials, such as insulation, windows, and appliances.

Green Seal is a non-profit organization, established in 1989, which helps consumers find truly green products through the Green Seal certification. This certification process is designed to ensure a product meets a set of multiple performance, health, and environmental criteria, also known as the Green Seal Standards. Examples of certified Green Seal products include cleaning supplies, construction materials, paint, paper, food packaging, and soap.

The *Programme for the Endorsement of Forest Certification* (PEFC) is the largest international forest certification system. The PEFC recognizes sustainability benchmarks in the forest and across the supply chain to ensure that forest products are sourced with respect for the highest ecological, social, and ethical standards. In the U.S., the PEFC has endorsed two certification standards, the Sustainable Forestry Initiative (SFI) and the American Tree Farm System. By labels or stamps on the wood products, consumers are able to identify and choose environmentally responsible products. Today, the PEFC has recognized certification in 30 countries and accounts for approximately 600 million acres of certified forests.

Environmental project permitting

In addition to what has been discussed thus far, numerous authorities having jurisdiction (AHJs), including many state and local agencies, have instituted a variety of environmental requirements and conditions. The construction team must plan, estimate, incorporate into their schedules, and implement these accordingly. Depending upon the location of the construction project site, there may be multiple environmental permits and restrictions placed on construction operations, regarding issues such as:

- Noise restrictions governing work days and hours,
- Parking restrictions for craftsmen,
- Traffic flow,
- Wetland protection,
- Storm water control,
- Demolition and abatement, and
- Hazardous construction material control.

An early submittal required by the project owner may be a project-specific storm water pollution prevention plan (SWPPP). This is similar to the safety and quality control plans discussed in previous chapters. This plan identifies potential sources of storm water pollution on the construction site and identifies measures to be implemented to eliminate polluted or silt-laden storm water from leaving the site. This often means taking steps to capture the storm water or retain it to enable infiltration into the soil. The SWPPP relies on erosion control as a primary means of preventing storm water pollution. Mats, geotextiles, and erosion control blankets may be used. Sediment controls provide a necessary secondary means of controlling storm water pollution runoff. Silt fences are often used as sediment control measures. The plan should address measures to be taken to control storm water flowing onto and through the project site, stabilize soils on site, protect storm drain inlets, and retain sediment on site.

The construction project may include demolition involving the removal of materials that contain *hazardous waste*, such as lead-based paint or asbestos. Proper documentation and disposal requirements need to be understood by all parties involved in the removal of the hazardous waste. Spill prevention plans are often required to reduce the potential for contaminating the soil as a result of construction operations, such as fueling equipment. Many hazardous materials may be used in construction operations, such as oil, solvents, paint, and glue, and any excess hazardous materials must be disposed of properly.

Project engineering applications

As mentioned earlier in this chapter, it is the project owner's responsibility to state their sustainability goals among the project objectives and write contractual documents that clearly convey them. A contractor's team, including construction project engineers, would need to know these requirements and use them as guidelines for delivering the project. A bulk of the contractor duties relates to documenting the use of sustainable products and materials and the administration of the rating and certification systems. Therefore, many contractual documents incorporate specific clauses to convey these requirements. Thus, some of the American Institute of Architects (AIA) contract documents also provide a revised version to be used on sustainable projects. For instance, a revised version of the standard A101 contract agreement based on stipulated sum pricing is available for use on sustainable projects and is called A101 SP. This revised contract document incorporates by reference a new document named the *Sustainability Plan*, which is supposed to outline all the sustainability requirements of the project owner.

As early as pre-construction planning, the general contractor (GC) should consider sustainability requirements to achieve a specific certification in the selection of materials, subcontractors, and construction strategies. Material reuse minimizes construction waste and earns credits toward certification. Documentation is needed to demonstrate the achievement of certain credits, and the submission of needed documentation must be included in supply contracts and subcontracts. A project engineer is often assigned the responsibility of collecting the documentation and assembling the sustainable materials needed for submission to a third-party certification agent to validate achievement of necessary sustainability credits. Sometimes this individual will also be designated as a "sustainability engineer," primarily because he or she has been accredited with the specific rating and certification systems, such as LEED. The estimated cost of the project engineer's time to manage the sustainability documentation needs to be included in the general contractor's general conditions cost estimate. A LEED requirement for building commissioning may be included with the contract close-out process. Close-out will be discussed in Chapter 14. As part of the material submittal process described in Chapter 11, the GC's project engineer is usually required to provide the following:

- Site use plan,
- Waste management and recycling plan,
- Indoor air quality plan,
- Identification of salvaged, refurbished and reused materials,
- Recycled content in materials,
- Regional material sourcing, and
- Certified wood products.

Summary

When the concept of sustainability appeared in the built environment, contractors viewed it as burdensome. Now, it just makes sense, not only in construction but in everyday life. Formerly, the pursuit of sustainability on a project cost more, but today new materials and improved building methods mean that it is not only reasonable to be sustainable, but it is actually more economical to operate and maintain a sustainable facility or infrastructure over its life. Sometimes, design professionals and contractors will incorporate green products and practices in new buildings and infrastructures, even if it is not an owner-generated requirement. Many municipalities are also adopting sustainable requirements in their building codes, so in the future, it will not even be an option to not go sustainable.

Review questions

1. What does the abbreviation LEED stand for?
2. What materials used in the construction process might be recycled?
3. Other than the engineered lumber products noted above, what are some examples you know of that are utilized in the construction process that help reduce our carbon footprint?
4. What three elements make up the triple bottom line? Which of these three is most important to design professionals or contractors? Which is most important to an owner? Which is most important to you?
5. If you were an architect or engineer, what green components might you recommend to your client?
6. Define "renewable resources."
7. What would happen if you eliminated one leg from the triple-bottom-line stool depicted in Figure 13.1?

References

EPA. (2017, April). Fast Facts from the Inventory of U.S. Greenhouse Gas Emissions and Sinks: 1990–2015. Retrieved September 27, 2017, from www.epa.gov/sites/production/files/2017-04/documents/fastfacts_20170413-11am_508.pdf

U.S. Energy Information Administration (EIA). (2017, September 26). Table 5.1: Sales of Electricity to Ultimate Customers. Retrieved September 27, 2017, from www.eia.gov/electricity/monthly/epm_table_grapher.cfm?t=epmt_5_01

LEED. (2009). *Illustrated Green Associate (GA) Study Guide* (2nd ed). Studio 4, LLC.

UN. (1987, August 4). Our Common Future. Report of the United Nation (UN) World Commission on Environment and Development. Retrieved September 27, 2017, from https://sustainabledevelopment.un.org/milestones/wced

14 Close-out

Introduction

As the physical construction of a project nears completion, contractors will transition into a close-out phase. The jobsite team should develop a project close-out plan to manage the numerous activities involved in closing out the job. Just as start-up activities are essential when initiating work on a project, good close-out procedures are essential to timely completion of contractual requirements, receipt of final payment, and release of retention funds. Some of the reasons the general contractor (GC) should pursue a timely close-out include:

- To officially end the clock on potential liquidated damages;
- To flush-out late change order proposals by subcontractors and close the door for future claims: Initiation of the close-out process forces subcontractors to submit their claims;
- To begin the clock on the warranty: Most contracts include a fixed-term warranty, so the sooner a project is closed out, the sooner the warranty will begin and end;
- To maintain good relations with the project owner, designer, and subcontractors: Since the GC may need to work with these firms again on future projects, good relationships will be important to secure and to successfully carry out future projects;
- To minimize jobsite overhead costs: The close-out process ideally requires involvement by the project manager, the superintendent, the project engineers, and the foremen;
- To receive the final progress payment: Although the last monthly progress payment should not be tied to close-out, some clients may hold it together with retention funds as an incentive for timely close-out;
- To close out the subcontracts and purchase orders as soon as possible: The project manager and project engineer must work closely with the subcontractors and suppliers to get the project closed out so that all parties may receive their retention; and
- To receive release of the retention.

An efficient project close-out is not only good for the general contractor, but it is also good for the project owner. All contractors want to close out a job quickly and move on to another project. Minimizing the duration of close-out activities generally enhances profit, because it limits jobsite general conditions costs and facilitates timely receipt of the final payment and release of the retention. Use of the project manager and superintendent to oversee project close-out can be expensive, and some contractors rely on less-expensive project engineers and foremen to manage close-out. Still, an inefficient close-out may inflate the costs and impact the GC's bottom line on the project. Moreover, the retention held may be approximately equal to the fee on a typical commercial project; therefore, the GC and all its subcontractors cannot realize profits on their projects until the retention has been collected.

   Efficient close-out and turnover procedures also minimize the contractor's interference with the project owner's move-in and start-up activities. A contractor may lose credibility with its clients because of inefficient close-out procedures. An unhappy project owner will probably be a lost client, because they may not give a contractor an opportunity to secure future projects if the current one was not properly closed out.

Close-out planning

The close-out planning process should begin once the contract is signed, if not when the project was initially considered for bid. The time it takes to properly close out the project and receive the final retention can be considerable. If contractual requirements for the final close-out are deemed excessive, they could financially damage the general contractor or its subcontractors. Thus, it is better to address these risks early rather than later. For example, if early or partial releases of the retention are going to be requested by the GC and/or its subcontractors, they need to be negotiated into the prime contract agreement upfront.

When the project is being bought-out, as discussed in other parts of this textbook, the management team should understand the close-out requirements for each of their subcontractors and clearly describe those processes within the subcontract and purchase order agreements. Thus, the project manager (PM) and project engineer (PE) need to search the specifications for close-out requirements not only for their own direct work but for that of their subcontractors as well.

We use many document control logs in construction, and close-out is no exception. For example, a close-out log could be created early in the process by the project manager and shared with one of the project engineers. The engineer may then be tasked with tracking completion of the operation and maintenance (O&M) manual from the mechanical subcontractor, the warranty from the elevator manufacturer, the surplus ceramic tile material, the touch-up paint included on the punch list, the signed-off permits of the utility subcontractor, and the lien release from the shoring contractor. It is by use of a close-out log, similar to the one shown in Figure 14.1, that this process is properly managed at the jobsite level. The general contractor team should review the specification sections for each of the subcontractors and list all required close-out items. The log should be submitted to the project owner or their agent to request approval by utilizing the same process as for other submittals, as discussed in Chapter 11. Now is the time for the general contractor to get a clear understanding of what the project owner and their consultants expect at the time the project is scheduled to be closed out. The log should also be issued to the subcontractors and suppliers early to remind them of their contractual responsibilities. Some subcontractors may be slow at closing out their portions of the project, so ongoing reminders from the responsible project engineer are often necessary.

Close-out implementation

Construction close-out is defined as the process of completing all of the construction tasks and assembling all documentation required to close out the contract. The prime construction contract will address requirements for substantial completion and final payment. If mismanaged, close-out can take up to a year after the receipt of a certificate of occupancy, which results in financial stress to the general contractor and its subcontractors. In this section, we discuss various elements of close-out implementation.

Commissioning

Commissioning basically involves testing and re-testing most, if not all, the systems and components to ensure the project meets the owner's requirements. The purpose of this phase is to minimize surprises when the project owner assumes control and operates the completed

SUBCONTRACTOR AND SUPPLIER CLOSE-OUT LOG

Reliable Construction Company

Project Name: Rose MXD, Project Number 414   Project Manager: Jason Campbell   Updated: 10/19/19

Description	Shoring	Site Utilities	Supply Rebar	Plumbing	HVAC	Electrical	Drywall	Ceramic Tile	Vinyl Flooring	Paint	Elevator	Landscape
Punch complete	X	X	NA									
Turn in permits	X	X	NA		NA		NA	NA	NA	NA		
Extra materials	NA	NA	NA	NA	Filters	Lamps	NA				NA	NA
O&M manuals	NA	X	NA									
As-built drawings	By RCC	By RCC	NA				NA	NA	NA	NA	NA	NA
Final lien release	X	X	X									
Test certificates	NA	X	NA				NA	NA	NA	NA	X	NA
Union affidavits	X	X	NA				X				X	NA
Back charges clear	X	X	X				X				X	
Demobilized	X	X	NA				X				X	
Warranties	X	X	NA				X				X	
Retention released			NA									

Figure 14.1 Close-Out Log

facility themselves. A commissioning plan is developed during the design phase that outlines the commissioning process to be used. Primary commissioning activities include verification testing of selected equipment and systems, training of the owner's operating and maintenance personnel, and identification of any defects or warranty issues that may be realized in the initial years of operation. Commissioning does not take the place of normal start-up testing and balancing of the mechanical, electrical, and plumbing (MEP) systems. Instead, it is an additional step. MEP equipment and system commissioning tests are conducted by using procedures developed for each specific project based upon on the owner's requirements.

MEP systems undergo aggressive commissioning tests that are intended to duplicate the climatic seasons the building will experience. *Design days*—including extremely warm, cold, and humid days—are simulated as part of commissioning. The commissioning process may be performed by a third-party commissioning agent who will be contracted directly by the project owner. The cost of an independent, third-party commissioning agent may be in the range of one percent of the total construction cost. Many mechanical contractors also offer commissioning services as part of their scope of work. Some project owners who have experienced in-house facility managers may self-perform commissioning. The GC will often designate one project engineer to coordinate the commissioning efforts of its subcontractors. A separate specification section will define the roles and responsibilities of the commissioning team. Commissioning teams for large or complex projects, such as hospital projects, will likely include the following individuals and companies:

*   MEP equipment representatives,
*   MEP subcontractor PMs and/or superintendents,
*   General contractor's project manager or project engineer in charge of MEP coordination,
*   Mechanical and electrical engineers,
*   Owner's maintenance and operation personnel (end-users),
*   Owner's representative, and
*   Third-party commissioning agent, if specified.

Construction close-out

As was the case with physical mobilization, most of the work related to completing construction execution and closing out physical aspects of the project is the responsibility of the project superintendent. Some of the construction activities associated with closing out construction execution include:

*   Verifying construction completeness in accordance with contract requirements,
*   Developing the formal punch list and the resolution of outstanding items,
*   Obtaining certificates of substantial completion and occupancy,
*   Facilitating project move-in and occupancy, and
*   Demobilizing resources.

Formal punch list

A punch list is a list of issues that need to be addressed to guarantee completeness in accordance with contractual requirements. The superintendent, project engineers, and other quality control specialists inspect completed work and identify deficiencies throughout the construction of the project as part of an active quality management program, as discussed in Chapter 9. This ongoing process, which ultimately saves all parties money, produces interim punch lists that are one of the tools a GC will utilize to reduce the size of the final punch list. When the project is nearing completion,

PUNCH LIST

Reliable Construction Company

Project: <u>Rose MXD, Project 414</u>   Inspection Date: <u>October 24, 2018</u>

Area/System: <u>Apartment Unit 304</u>   Status Date: <u>October 30, 2018</u>

Inspection Participants: Ted Warren/Owner, Robert Jackson/Arch, Susan Thompson/Superintendent, David Arnold/QC, Jennifer Forsythe/PE

Item	Description	Complete?	Verified?
1	Vinyl plank floors edge curling	X	ST
2	Paint touchup throughout		
3	Refrigerator ice maker not connected	X	
4	Light bulb dim in bedroom 2		
5	Master bath faucet drip	X	ST
6	Missing shower rod	X	ST
7	Small chip in ceramic tile, powder room	X	ST

Figure 14.2 Punch List

the GC requests a *walkthrough inspection* from the project owner and their consultants. The list of deficiencies identified during this inspection is known as the formal punch list. Figure 14.2 shows an example of a punch list. Whereas some project owners perform their own inspections to prepare the formal punch list, this task is usually delegated to an owner's consultant. On building projects, the architect or the agency construction manager are frequently tasked with this duty. On civil construction projects, this might be the project owner or civil engineer. Some consultant design team members, such as mechanical or electrical engineers, may also develop punch lists specific to their scope. Many separate groups within an owner's organization may also develop their own separate punch lists, depending upon their specialization. The best method would be for all parties who are interested in inspecting the project to walk the jobsite at the same time. One collective punch list should be developed at the end and issued to the general contractor. Alternatively, the owner and design teams may provide input to the GC's project manager or project engineers, who will then compile the issues into one official punch list to be shared with all parties.

The final punch list walkthrough should occur early enough to allow the GC and its subcontractors sufficient time to address the issues prior to the project owner taking occupancy. If the team is still developing the punch list after the project owner has moved in or assumed control, it is difficult to determine who did what damage. However, it is a mistake to start the process too early. If there are still basic construction activities to be completed, additional trade damage can occur after the walkthrough that would not have been listed on the original punch list. Therefore, the formal punch list should be developed when all major construction activities are completed.

A realistic goal for the GC is to take no more than one month to complete all the work on the punch list. All items on the punch list must be corrected before the project can be considered complete. If the process takes too long, the responsible parties will likely have demobilized, and it may be difficult to get them back to the jobsite. The punch list should be signed off by each subcontractor and the GC as deficiencies are corrected. A copy of the annotated punch list should be sent to the owner's consultant, who is usually the architect on building projects. The architect may then wish to re-visit each punch list item and verify its completion or may perform spot-checks on items or rooms. The architect will then notify the project owner that the punch list has been corrected and the certificate of substantial completion can be issued.

Certificates of completion

There are two certificates of completion needed for contract close-out. The *certificate of substantial completion* is usually issued by the architect on commercial projects and the civil engineer on heavy-civil public projects. This important certificate indicates the project is sufficiently completed such that it can be used for its intended purpose, but there still may be some minor deficiencies that need to be addressed, hence "substantial." Although all items on the punch list may not have been corrected, the project owner or their consultant agrees that the project is ready to be used. A list of the pending issues in the form of an updated punch list should be attached to this certificate. Receipt of the substantial completion certificate is a significant milestone event; it ends the GC's liability for liquidated damages and starts the clock on the warranty period. Contractual completion is the date by which substantial completion is to be achieved.

The *certificate of occupancy* (C of O) is issued by the authority having jurisdiction (AHJ) over the project site, which is likely the city, county, state, or federal government. This is the same agency that issued the original building permit. This certificate signifies all code-related issues have been accounted for and that the facility is approved and safe to use. It is often a formality following completion of all the various inspections and the subsequent approval and sign-off of all other construction permits. The inspection and approval from the fire marshal for life-safety issues is usually the most critical step for the issuance of the certificate of occupancy. The project may have minor deficiencies that are not related to life-safety and still be approved for occupancy. For example, the ceramic tile base in apartment # 304 may not yet be complete or corrected, but the apartment is still usable. Some public agencies will issue a *temporary certificate of occupancy* (TCO), which allows conditional use of the building for a stated period of time—say six months—while other non-critical work is being completed, such as landscaping or some interior finishes. The GC needs to obtain both the certificate of substantial completion and the certificate of occupancy to signify that the project is complete.

Occupancy

Most project owners want to move into their new building as soon as the AHJ allows it to be occupied. Joint occupancy occurs when the project owner accepts and occupies a portion of the facility, or takes over control, while the contractor is still working in another portion. This may be dictated by the owner's need to begin business in the new facility, lease out apartments, or begin moving equipment and furniture into the building. Whereas joint, dual, or conditional occupancies are often unavoidable, they may become detrimental if not managed properly. The project team may have problems differentiating the client's cleanup and routine maintenance from the construction punch list and warranty work. It is better for all parties if occupancy is delayed until both certificates of completion have been obtained and all items on the punch list have been corrected. If the GC and the project owner are to have joint occupancy, both parties must work together to establish procedures to accomplish each other's goals. The project owner's right to take partial occupancy of an uncompleted project should be described in the prime contract.

Demobilization

Similarly to mobilization, most building estimators do not always put a line item in their original estimate for demobilization. Heavy-civil projects have significant mobilization and demobilization estimate line items. Demobilizing, or physically moving off the site, involves considerable work and can be expensive, especially to the general contractor. At that time, there may not be funds that can be dedicated to a foreman and small crew to clean up the site. The question always arises between the GC and its subcontractors as to whose garbage remains. As with the discussion regarding joint occupancy, garbage accumulation may become mixed between the construction crews and the project owner, if the project owner is also in the process of moving in. Demobilization involves closing the project office and removing all contractor-owned equipment from the project site. The project staff is phased out and reassigned to other projects, and record files are taken to the GC's home office.

Contractual close-out

Contract close-out begins with checking the specifications for close-out requirements. Ideally this happens during project start-up, so that the requirements are delineated in each subcontract and major purchase order. The project manager or project engineer should make a list of what they think needs to be done and then ask the design team for verification. This list provides a basis for developing a close-out log, which is similar to an early submittal log. Some of the major items that the project team will submit during the close-out process include as-built drawings, operation and maintenance manuals, test reports, extra finish materials, and signed-off permits.

As-built drawings

The development of *as-built drawings* is a substantial and important close-out process. Actual dimensions and conditions of the installed work are noted on the contract drawings. The project manager and jobsite project engineer should collect all of the as-built drawings from the subcontractors, according to what is required by the owner in the specifications. Mechanical, electrical, and civil installations are very important, because they include hidden systems that would cause severe damage if cut or that may need to be accessed during a future remodel or phase. Usually the general contractor will develop as-built drawings for the architectural and structural work; MEP subcontractors will prepare their own.

The most appropriate project team member to mark up the as-built drawings is the foreman or assistant superintendent who oversaw the work, but this job may sometimes be delegated to the project engineer. The as-built drawings should be submitted to the project owner or designer by using procedures similar to those used for shop drawings. Projects relying on building information modeling (BIM) or computer-aided design (CAD) software for the design process may also require contractors to record as-built conditions electronically. If electronic as-built drawings are required, an additional specialized PE is often required on the GC's team and must be included in the jobsite general conditions estimate.

Operation and maintenance manuals

O&M manuals are large physical volumes that gather manufacturers' data regarding operational, preventive maintenance, and repair procedures for all equipment, as well as many of the architectural finish materials. This information should be collected from all subcontractors and suppliers and organized in several sets of three-ring binders for the project owner's permanent service record. Often, the contract specifications will dictate the format and organization of these documents.

Processing the O&M manuals as a submittal and requesting designer approval is sometimes required and is always a good idea. Some project owners will request draft copies be submitted for comments prior to submission of the final copy. The project team should not simply collect previous submittal data and place it into a three-ring binder. Similarly to as-built drawings, electronic submission of the O&M manuals may be required. These manuals should include information that the project owner can use for its operations and maintenance, not marketing propaganda. The

GC should provide a service to its client, the project owner, and ensure their subcontractor O&M manuals fully conform to contract specifications.

Test reports

Reports are created during the course of the project as materials and systems are tested. This includes balancing reports for the mechanical systems. All of these test reports, along with commissioning documentation, may be bound as a section in the O&M manuals. Collection of these reports and diligent filing throughout the course of the project by the project engineer will help with final assembly of the O&M manuals.

Extra materials

Specifications often require the general contractor to supply extra material quantities to the project owner upon completion to assist with future repairs and remodels. For instance, one to three percent of various architectural materials may be requested; this may include materials such as paint (of each color used), ceramic tiles, carpet, and ceiling tiles. Management of these extra supplies is important because construction craftsmen sometimes incorrectly use up these materials for punch list or change order work and contractors risk being short of materials at the time of turnover. Ways to prevent this are to either lock the materials up or turn them over as they become available and have the project owner sign for receipt.

Permits

Each subcontractor that was required to provide its own permit from the AHJ also needs to get it signed off with a final approval from the AHJ. This often includes firms such as electrical, elevator, and fire protection contractors. This indicates that the subcontractor's work was performed in conformance with code requirements. The jobsite management team may also be required to submit to the designer all interim inspection reports or signature cards received from the city or county throughout the course of the job. Similarly with many other close-out documents, signed-off permits may be included as a section in the O&M manuals.

Financial close-out

The GC's project manager is also responsible for financially closing out the project with both the project owner and the GC's subcontractors. Upper management and project engineers may assist in this task. The financial close-out of a project includes several tasks, including negotiating final change orders with the project owner and, in return, negotiating final change orders with the subcontractors and suppliers. Additionally, financial close-out includes receipt of final pay requests from subcontractors and, in turn, issuing of the final pay request to the project owner. Once paid by the project owner, the GC's project team should expeditiously release payments to subcontractors. The pay request process will be accompanied by final lien releases from all of the GC's vendors and preparation of the same for the project owner. Many of these tasks are elaborated on in Chapter 22.

In-house close-out

There are many additional in-house close-out activities that are the responsibility of the project team. Some of these include:

*   The *as-built schedule* is prepared from actual dates and durations marked up on the jobsite meeting room schedule. The as-built schedule records actual deliveries and actual activity start and completion dates;

- The *as-built estimate* should be maintained throughout construction or prepared near completion of the project but not after close-out. Considerable work goes into tracking actual costs. This is valuable input to the construction firm's ongoing ability to improve its estimating accuracy. Input of the as–built estimate into the estimating database is necessary if the database is to be kept current. Many construction managers will simply input actual costs alongside the original quantities in the company's database. This is better than no input at all, but the most accurate historical cost data are created by combining actual direct labor hours and actual material costs with actual installed quantities.

Example

If actual material costs and hours are not factored in with the actual installed quantities, the historical database would be less useful in estimating future projects. Portions of the quantity take-offs (QTOs) for the Rose and I-90 cases described in Chapter 6 would incorporate actual materials costs and hours as follows:

- *Rose Case:*
  Estimated: 515 door leafs and hardware sets @ 6 MH/EA = 3,090 MHs
  3,090 MHs @ $29/HR = $89,610 labor cost
  Spent: $82,800 to install 515 door leafs and hardware sets; therefore:
  $82,800/$29/HR = 2,855 hours/515 leafs = 5.54 MH/EA labor productivity
- *I-90 Case:*
  Estimated: 120 EA ABs @ $18.10/EA = $2,172 material cost
  Spent: $2,620 to purchase 130 ABs; therefore:
  $2,620/130 = $20.15/AB unit price to purchase

Warranty management

The project owner is further assured that the project was completed in accordance with all contract requirements by the issuance of a warranty from the GC. The warranty period specified in most construction contracts for the entire project is one year after issuance of substantial completion. The warranty countdown could start earlier on portions of the project occupied by the owner prior to substantial completion if allowed by the contract. Longer warranties may also be required in the technical specifications for selected components, such as roofing, glazing, elevators, or an emergency generator.

  In terms of warranty, general contractors should incorporate in their subcontracts the same contractual requirements as those stated in their prime contract with the project owner. Two aspects of warranty management should be clearly conveyed in the subcontracts. First, the warranty period should match that of the prime contract because the general contractor will rely on the subcontractor that performed the warranted work. Second, collection of subcontractor and supplier warranties is another element of the project close-out process and should be clearly stated in the subcontracts. One of the authors experienced at first hand a failure in implementing this common-sense practice when he, as a young project engineer, and his team were tasked with gathering warranty documentation for electrical panels installed at more than a hundred cell phone radio stations. This last-minute gathering task was needed because the electrical subcontract did not require the electrical subcontractors to submit this documentation in binders, so they left it in the electrical rooms of each site. However, the prime contractor was required to submit a full documentation binder to the project owner, a cell phone operator, as part of the close-out. When subcontracts are well drafted, they address this issue, so the project manager and project engineers can simply request written, original, and signed warranties and guarantees from all subcontractors and suppliers. All warranties should be submitted by subcontractors and suppliers on their own letterhead. Warranty submission should be monitored and tracked in the close-out log, as shown in Figure 14.1. Final payments to subcontractors and suppliers should not be made until warranty documents have been received. The warranties can all be inserted as another section in the O&M manuals. If possible, the GC should keep subcontractor performance bonds active until the end of the warranty period to protect against any subcontractor failure to respond to warranty claims.

Support for the project owner during the warranty period is also a project management function, which may be performed by the project manager, superintendent, or project engineer. Warranty service basically involves some of the following:

- Receiving warranty claims from the project owner,
- Logging warranty claims (similar to other document control logs discussed elsewhere),
- Responding to claims with the GC's direct craft forces or subcontractors,
- Receiving the project owner's approval that the issue has been resolved and noting it on the warranty log, and
- Walking the project with the project owner just before the end of the warranty period to verify that there are no additional outstanding issues.

Project engineering applications

A jobsite project engineer often plays a critical role in the close-out process by assisting the project manager and the superintendent in a variety of activities. She or he will utilize many of the document processes and technology tools discussed in previous chapters and apply them to the project close-out activities as well. Some of the PE's close-out activities would include the following:

- Verifying subcontractor close-out requirements are included in their contracts;
- Developing and maintaining the close-out log;
- Performing interim punch lists;
- Assisting with development of the formal punch list and expediting its resolution;
- Collecting various documents and reports to be included in the O&M manuals, such as test reports, signed-off permits, and warranties;
- Assisting with development and collection of as-built drawings and O&M manuals, and submitting and receiving approval from the project owner and design team;
- Expediting receipt of final change orders, final pay requests, and final lien releases from subcontractors; and
- Assisting or providing input in the development of the as-built estimate and as-built schedule.

Young project engineers may overlook this process in their eagerness to move forward to a new project. Participation in the close-out process should not be looked upon by the jobsite project engineer as degrading, but rather as empowering. More so than with many of our other construction management topics, this is an opportunity for a project engineer to take a more major role in jobsite construction management. Often the project manager and superintendent have demobilized and started a new construction project, and it remains for a project engineer and a foreman to efficiently close out the project. This is an opportunity to build long-lasting relationships with other contractual parties. The reasons for a client to appreciate a prompt close-out should be clear by now. The subcontractors appreciate timely release of their retention. Moreover, the GC's home-office management team will recognize the project engineer for successfully performing the tasks necessary to receive their final retention and fee. A well-managed close-out phase often results with a seasoned project engineer becoming a project manager for this same client on their next project. In this way, close-out should be seen as a "get-to" and not a "have-to" for a project engineer.

Summary

Construction close-out involves completing all of the construction tasks and processing all of the documentation required to close out the contract and to consider the project complete. The project manager, project engineer, and superintendent work closely together to ensure close-out procedures are comprehensive and efficient. Good close-out procedures typically result in higher contractor profits and satisfied clients. The management of close-out documentation is the responsibility of the project manager and project engineer, as physical completion of construction is the responsibility of the superintendent. The project engineer will utilize many of the technology tools discussed in prior chapters for close-out, in addition to modeling tools introduced in the upcoming chapter.

As part of active quality management, project engineers and foremen perform walkthroughs as tasks near completion and develop interim punch lists for subcontractors. As the project nears completion, the GC will schedule a formal punch list walkthrough with the project owner and/or its consultants. Any deficiencies noted during this inspection are placed on the formal punch list for future re-inspection. All deficiencies on the formal punch list must be corrected before the contract can be closed out. A significant project milestone is achieving substantial completion, which indicates that the project can be used for its intended purpose. On building projects, the architect decides when the project is substantially complete and issues a certificate of substantial completion. However, the project owner cannot move in to their new project until a certificate of occupancy has also been issued by the city, county, state, or federal government.

Financial and contractual close-out of the project is the GC's project manager's responsibility with the support of his or her project engineer. This involves issuing final change orders to subcontractors and major suppliers and securing their final and unconditional lien releases. As-built drawings, O&M manuals, warranties, and test reports must be assembled. The project manager or project engineer should develop a close-out log early in the project to manage the timely submission of all close-out documents. An efficient close-out of all project activities allows the contractor to receive its final payment and release of retention, which is its ultimate goal.

Contractors are typically required to repair any defective work or replace any defective equipment identified by the project owner within the one-year warranty after substantial completion. Warranty response is an important aspect of customer service. Poor warranty response on a project may jeopardize the contractor's ability to obtain future projects from the same project owner.

Review questions

1. What items are tracked on a close-out log?
2. When does the warranty period begin?
3. From whom should the GC obtain written warranties?
4. Where would you determine which components require warranties that are longer than one year?

5. Why is warranty response an important aspect of customer service?
6. Where does the project engineer obtain the information for the requirements of the O&M manuals?
7. What is an interim punch list?
8. When should the formal punch list be developed?
9. Why is substantial completion a significant contractual milestone?
10. Who issues the certificate of substantial completion?
11. Who issues a certificate of occupancy?

Exercises

1. Prepare a close-out log for a project other than our case studies. Include at least ten subcontractor and supplier categories.
2. Assume you were the project engineer for the construction of a new hospital. The hospital administrator has made a warranty claim regarding the failure of a nurse call station. You have notified the electrical subcontractor, but the subcontractor has failed to take any action. What action should you take?
3. Draw a flow chart of a material procurement process (pick one from either of our case studies) starting at the selection of the supplier and ending with receipt of the product warranty. Which of these activities would be a "get-to" for the project engineer? You will also need to refer to other chapters in the textbook for this exercise.

15 Modeling project documents

Introduction

Throughout this book, various project documents have been introduced. In Chapter 5, construction contractual documents were discussed, including the *contractual agreement* that incorporates by reference other contractual documents, such as the *general conditions, special conditions, construction drawings*, and *construction specifications*. In later chapters, other documents that are used to manage construction processes according to the contractual documents were introduced, including *project schedules, project estimates, requests for information* (RFIs), and *change orders*.

Management of a project out of all these documents is cumbersome and may result in information breakdowns that can negatively affect a project's outcome. As a result, project team members from both the design and construction disciplines have continually experimented with approaches to streamline and visualize the information flow. In the pre-digital era, streamlining efforts produced the use of project logs and scaled-down models of the project, including scaled architectural models, scaled models of architectural details, and mock-ups. Over the last decades, digital tools have provided additional opportunities and challenges for both designers and contractors. This chapter will provide a brief overview of some of these tools, including how they evolved over time.

To explain the differences between modeling tools, built environment projects will be categorized into vertical and horizontal projects. Within the scope of this chapter, *vertical projects* include buildings and other projects where the vertical dimension is comparable, if not predominant, against the other two. Therefore, a large amount of the design and construction information about vertical projects deals with vertical dimensions. Typical vertical projects include buildings and industrial facilities. On the other hand, *horizontal projects* are defined as those where the horizontal dimensions are predominant in the design and construction information. Infrastructures, such as roads and utilities are usually classified as horizontal projects. Three-dimensional (3D) digital modeling tools are usually prevalent on vertical projects. Conversely, horizontal projects often rely throughout their initial project phases on two-dimensional (2D) modeling tools and may introduce 3D information to a lesser extent only during the late design and construction phases.

From computer-aided design to full digital modeling of project documents

Before the mid-1980s, design firms highly relied on armies of drafters working in rooms full of drafting tables to produce drawings for built environment projects. In the late 1980s, 2D computer-aided design (CAD) software programs were slowly introduced to the design practice to automate the drafting process and the production of drawings. These programs allowed for an initial paradigm shift by producing the equivalent of paper drawings on a digital drafting table, where they could be easily modified, replicated, and resized to the desired scale before being sent to a plotter to produce a standard set of drawings for procurement and construction.

The main advantages of this approach were the ease of revising drawings after each meeting with the project owner and the increased efficiency of developing a single 2D model of the project that could be printed at various scales depending on the use. However, these programs were simply electronic drafting tables where the objects on the computer display were just sets of lines, so they did not affect the underlying design processes. Examples of these programs include early versions of AutoCAD by Autodesk, MicroStation by Bentley Systems, and CATIA by Dassault Systèmes.

Whereas these new tools were valuable to the design profession, they did not transform the traditional design process, in which facilities were envisioned through 3D sketches and then drawn as 2D drawings. A common thought among designers was, "Since we conceptualize our projects in three dimensions, why do we still need to draw them in a two-dimensional fashion?" In the 1990s, new 3D programs were made available. Initially, pre-existing 2D tools were adapted to allow 3D modeling and create solids shaped as columns and slabs, which led to new versions of AutoCAD, MicroStation, and CATIA. However, new object-oriented CAD tools also appeared that allowed the treatment of the typical components of built environment projects as parametric entities that could be created to populate libraries for use by designers, who could shape and adjust them to the project needs. Examples of these tools include ArchiCAD by Graphisoft and Allplan by Nemetschek. These object-oriented CAD tools incorporated parametric modeling that allowed the creation of more complex assemblies specific to vertical projects, such as windows, composite walls, and roofs.

This second generation of CAD tools allowed designers to create a 3D digital model of the expected final product that could be observed from various viewpoints to evaluate its shape and function and identify potential conflicts between building elements at the macro level. Once approved, the 3D model could be used to generate 2D drawings for procuring and contracting construction services. These tools also promoted some initial changes in the underlying design processes. Object-oriented tools allowed automatic counting of assemblies and volumes, which provided for rudimentary quantity take-offs. However, several steps of the modeling process were still not automated, and designers still depended upon other computer tools for structural and mechanical engineering analyses. Therefore, the initial armies of manual drafters and rooms of drafting tables were simply replaced by smaller armies of computer drafters and smaller rooms of computer stations.

Overall, the widespread adoption of CAD tools by design firms introduced digital modeling and transformed the production of drawings, but it did not significantly affect other construction documents. In the 1990s and early 2000s, project delivery still relied highly on separated contracting for design and construction services. As such, on the construction side, the adoption of CAD tools by designers had a limited effect on contractors, because they were still selected based on paper documents and expected to abide by them. Moreover, paper drawings were easier to use in the field than the bulky technology available at the time.

Today, technology is bringing CAD information into the field through the miniaturization of computing equipment and development of advanced communication technologies. For instance, tablets allow project engineers, superintendents, and foremen to access the latest set of project drawings. They also allow builders to attach photos or videos to record as-built conditions. Moreover, it is now possible to print 3D scaled models of the building or its components that can be used to communicate with other project team members, as shown in Figure 15.1.

From an information technology viewpoint, the widespread adoption of CAD tools made it possible to think more boldly, and it initiated a push toward the full integration of multiple computer tools for estimating, scheduling, and planning into a single family of tools that operated

Figure 15.1 Use of 3D Printing for Project Communications
(Credits: Matt Glassman, PCL Construction)

around a geometric model of the built environment project. This new generation of tools does not simply support existing design processes—it has radically transformed both design and construction processes through integration and virtualization. Two families of virtual design and construction (VDC) tools have been developed in support of vertical and horizontal projects, namely the *building information modeling* (BIM) and *civil information modeling* (CIM) tools that will be discussed in the rest of this chapter. Differently from CAD tools, which were intended as standalone programs, BIM/CIM tools are modular. Each major provider of BIM/CIM tools relies on its latest generation of CAD tools to serve as the core for a wider family of tools that interconnect with each other to support most design and construction processes. Whereas 2D drawings are still prevalent in contracting, this technological advance is also introducing the use of 3D models as an additional contractual document.

Because project teams still rely on large amounts of 2D drawings, design and construction firms are also using other simplified tools that allow the markup of 2D drawings and hyperlinking of the drawings with textual documents. This is a way of enhancing communications and collaboration between team members. Whereas these tools do not model project documents, they have become prevalent in the practice of a construction project engineer. Therefore, a section of this chapter is dedicated to markup and collaboration software. Some of these tools include Bluebeam Revu by Nemetschek, AutoCAD Design Review by Autodesk, and PlanGrid by PlanGrid, Inc.

Building information modeling

One of the earliest object-oriented CAD tools, ArchiCAD by Graphisoft, could be credited as the precursor of building information modeling, with its virtual building concept already implemented in the mid-1980s. BIM is not simply a family of computer tools but a new approach toward the integration of design and construction processes. At this time, BIM is one of the buzzwords in the delivery of vertical projects. Thus, there is much to say and to learn about BIM. The intent of this section is to provide an overview of what BIM is and how it is used at different phases of a project. The recent widespread adoption of BIM tools is due to both the technological advance of CAD systems and the adoption of project delivery processes that integrate the contracting of design and construction services. To a different extent, all BIM tools are designed to promote integration between the geometric model of the intended outcome, processes needed to produce a fully built and operational project, and various contractual documents. Therefore, a BIM model goes beyond the three traditional geometric dimensions of 3D CAD systems to also include information such as unit costs, times of installation, and forecasted energy performance, among many others.

BIM during planning

As discussed in Chapter 2, a project concept is developed during the planning phase with a plan for delivering the project. As a result, there is little design to be performed at this time, which means there is little need for BIM tools. At this time, however, the project owner needs to make a few important decisions that are related to BIM, including the project delivery method to be used, if BIM tools should be mandated on the project, and what BIM deliverable should be produced.

BIM tools are integrated, so their implementation is easier on projects where design and construction services are integrated. On these projects, the benefits from BIM implementation can be maximized. However, not all designers and contractors use BIM tools, and if they use them, they may be using different products. Therefore, making the BIM model one of the deliverables for a project is an additional contractual requirement, which implicitly mandates the use of BIM tools and may affect procurement and delivery. This requirement would allow the owner to receive a comprehensive as-built model of the project, together with the physical facility, which will make it easier to implement maintenance and retrofits during operations. However, several tools exist, so it is important for the project owner to evaluate the advantages and disadvantages of requiring interested designers and contractors to use specific BIM tools versus using those that they prefer. Even those firms that utilize BIM tools daily in their own practice may incur additional costs to elevate the quality of their BIM model up to the project owner's contractual expectations. These costs would further increase if the project owner is requiring the design team to prepare the BIM model with a different computer program. To address this new type of contractual requirement, sets of standardized contractual documents now include BIM exhibits and addenda documents to be incorporated by reference in the original contractual agreement, including the American Institute of Architects (AIA) E202 and E203, the ConsensusDocs 301, and the Design–Build Institute of America (DBIA) E-BIMWD. These standard documents guide the project owner in making the right decisions about BIM requirements before contracting services.

BIM during design

The core of each BIM platform is usually based on a 3D CAD system, which is mainly a geometric design tool. In a BIM platform, the geometric component, however, is "connected" with other information contained in databases and other software tools that perform several necessary analyses and automate the information exchange. Depending on the selected BIM platform, different opportunities are available to connect internally within the platform of tools and externally with other computer tools. For instance, one of the most popular BIM platforms, Revit by Autodesk, includes tools that can be used for designing, fabricating, and installing architectural components, as well as building systems and structural elements. These tools are able to easily interact with each other within the Revit platform by using Autodesk's proprietary language. Moreover, Revit provides the opportunity to communicate with external tools through various formats for interchanging information, including the popular Industry Foundation Classes (IFC). This ability to communicate externally with other computer tools is usually defined as *interoperability*, and it is a crucial aspect to consider when selecting a BIM platform or evaluating the project owner's requirements on BIM. If a project owner is requiring their team to submit a BIM model in a format that is barely compatible with the platform used by the designer or contractor, this means they may need to hire a specialty consultant to recreate the model in the required platform.

BIM during construction

During pre-construction and construction, BIM tools provide several opportunities to support general and specialty contractors in their construction planning and execution. New tools are continually developed to improve cost, time, safety, and quality performance, so this section is only minimally illustrative of what BIM tools can offer to contractors.

For instance, four-dimensional (4D) BIM tools allow contractors to add a time dimension to 3D building components to simulate how they will come together over time. As part of these simulations, temporary construction elements, including scaffolding, formwork, site facilities, and equipment, can also be modeled to evaluate any physical constraint of the site, issues with site logistics, and conflicts between the facility being built and overhanging utilities and cranes; this allows the contractor to forecast potential safety issues.

Moreover, mechanical, electrical, and plumbing (MEP) firms model their systems with BIM and utilize clash detection tools to verify potential conflicts between systems, as shown in Figure 15.2. For instance, BIM can be used to verify if conflicts may exist between the mechanical ducts and the structural, plumbing, or electrical systems during installation. Clash detection exercises allow for early identification and resolution of these conflicts, which minimizes the risk of field rework and allows for more certainty in cost and time forecasts. The same approach can be used for retrofit projects, where the current situation can be digitalized through laser scanning, inputted into the model, and used to verify MEP elements will fit before fabricating them. Figure 15.3 provides an example of how these modeling efforts are visualized.

Other BIM tools allow for concrete volume estimation and rebar placement, with similar advantages to the clash detection for MEP systems. For instance, concrete modeling tools can be used to plan concrete lifts on high-rise projects, as shown in Figure 15.4.

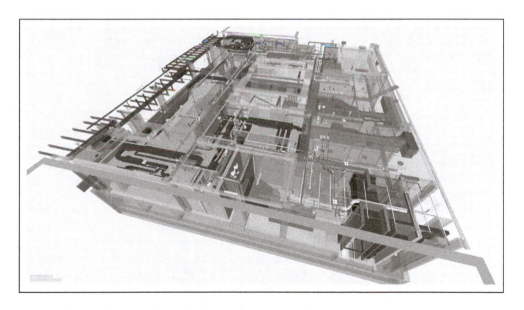

Figure 15.2 BIM Model for MEP Coordination
(Credits: Matt Glassman, PCL Construction)

Figure 15.3 Laser Scanning to Verify Compatibility of Designed MEP against Existing Conditions
(Credits: Matt Glassman, PCL Construction)

BIM models can also be used to support cost estimating efforts for contracts and change orders. Whereas estimating heavily relies on human intelligence for pricing, BIM tools can automatically produce quantity take-offs (QTOs) that can be used as ballpark numbers to verify that human-developed QTOs did not miss anything. It is not difficult to believe that the time for fully automated QTOs is quickly approaching. However, most projects still rely on 2D drawings for pricing purposes. Even when designers share their models with contractors, these models either come with disclaimers that shift the risk from using them to the contractor or they are not detailed enough to produce a complete QTO. Therefore, it is common for the contractor to develop their own 3D model before linking information from project documents.

BIM during operations

One of the main reasons for project owners to request a BIM model as a deliverable is to use and update the model throughout operations of the facility. BIM models for operations act as living as-built models. Several efforts are underway to incorporate into such models the information and documentation for all equipment installed, pre-cover 360-degree photos, and laser scanning of the final product. These efforts intend to produce a rich digital model of the physical facility. This way, the owner's facility managers can click on a room and access information on materials and equipment installed, its age, its operating manuals, information on how to reorder it, if there is a need to replace it, and the location of electrical conduits or plumbing lines behind the gypsum wall board. One of the efforts to make possible information exchange along the life cycle of a building is the Construction Operations Building information exchange (COBie) specification developed by a group of governmental agencies, including the U.S. Army, Corps of Engineers (USACE), the National Aeronautics and Space Administration (NASA), and the National Institute of Building Sciences (NIBS). Some of the major software programs for operations and facility management already provide interoperability with BIM software, including Maximo and TRIRIGA by IBM Corporation.

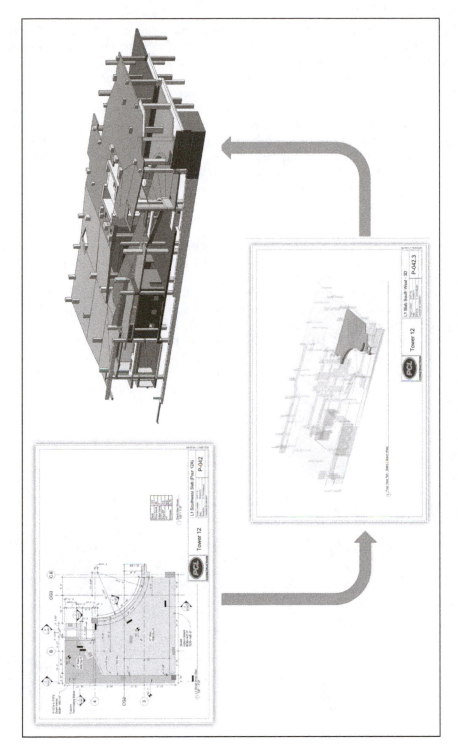

Figure 15.4 BIM Tools for Concrete Volume Estimation
(Credits: Matt Glassman, PCL Construction)

Civil information modeling

Horizontal projects, such as utilities and road infrastructures, mostly rely on two-dimensional geometries. To this end, different terms have been used to refer to modeling efforts for these projects, including civil information modeling, civil integrated management, and infrastructure BIM (infraBIM or iBIM). Currently, the term *civil information modeling* is the most used in the United States when referring to modeling tools for horizontal projects.

Whereas a predominant amount of geometric information in CIM models is two-dimensional, CIM also includes 3D tools designed to handle project aspects where the third dimension is important, including mass excavation models, bridge erection simulation, and visualization of proposals. CIM tools have often been inspired by BIM. However, the variety of these tools is usually less diverse than those of BIM. The rest of this section will focus on a few CIM features that are significantly different from the BIM features that have been discussed in the previous section.

Mass excavation models

The use of CIM to support earthmoving operations was motivated by another concurrent technological advance, namely, the use of construction equipment controlled through global positioning system (GPS) technology to implement automated machine guidance (AMG). Major earthmoving equipment manufacturers now offer AMG-enhanced equipment through GPS devices that allow contractors to automatically control the blades of dozers and graders to optimize the number of passes and speed of operation. However, the use of these hardware devices relies on the existence of a mass excavation model that can be imported into the equipment hardware. CIM allows development of these models. The joint use of CIM and AMG has allowed substantial improvement in performance through cost reduction and schedule certainty. Sometimes, these efforts are supported by other technologies, such as drone photography and laser scanning. Figure 15.5 shows

Figure 15.5 Laser Scanning to Evaluate Mass Excavation Volumes
(Credits: Matt Glassman, PCL Construction)

an example of how a major general contractor used laser scanning to survey the current status of a jobsite and evaluate volumes of excavation.

Integration with geographic information systems

Infrastructure projects usually develop over long distances so they need to be well integrated within their context. As a result, planning efforts rely on territorial information, which is often stored in geographic information systems (GIS) that allow public agencies to "link" information available in economic, demographic, road accident, and traffic databases to the location on 2D map files. GIS-stored information is usually used to guide planning and conceptual design efforts. Several efforts exist to automate the information exchange from GIS to CIM, and some public agencies have already developed protocol interfaces between the two systems.

Markup and collaboration software

BIM and CIM allow project teams to model project geometric information and integrate the models with other project documents. However, most projects still rely on two-dimensional drawings. To allow some level of integration of these drawings with other project documents, design and construction firms can use various tools that allow markup of two-dimensional drawings and hyperlinking of them with textual documents, as a way to enhance communications and collaboration between team members. Some of these tools include Bluebeam Revu by Nemetschek, AutoCAD Design Review by Autodesk, and PlanGrid by PlanGrid, Inc.

For instance, Bluebeam Revu allows users to create, markup, and organize project documents. It also facilitates collaboration among team members for revising and using project documents. Markups can be posted on documents, organized, and assigned to team members for action. A popular Bluebeam Revu feature allows estimators and project engineers to take off quantities directly from portable document format (PDF) files. Another Bluebeam Revu application allows updated sets of documents to be published with request for information (RFI) markups, as shown in Chapter 11, which details the electronic posting of RFIs to drawings. This approach allows contractors to maintain a centralized digital set of documents that is up-to-date. However, the effectiveness of this application relies highly on the project engineers to promptly post RFIs and other information on the right documents as soon possible.

Project engineering applications

Project engineers are highly involved in modeling project documents through BIM, CIM, and markup and collaboration software programs. Project engineers are usually the younger members in a project team, or at least the most recently graduated from an architecture, engineering, or construction program. Therefore, their project teams often rely on them to be aware of the latest technological tools, including those briefly described in this chapter. Independently from their college training, it is not uncommon for project engineers to receive further on-the-job training to master specific tools that they will be expected to use. In building projects, contractors often rely on a project engineer to update the BIM model. These specialized PEs are often called BIM engineers, and depending on the size of the project, they can be either project- or home-office-based. Because MEP coordinators are fully vested in overseeing and coordinating MEP specialty designers and contractors, they are also often delegated with the management of the BIM functions that pertain to the MEP systems, including clash detection simulations. Any other project engineer is expected to have enough knowledge of BIM or CIM to access information on the model and to acquire mastery of markup and collaboration tools as needed by the scope items they work on.

Summary

Projects rely highly on numerous project documents, including drawings. As the project evolves, these documents frequently change, which forces the project team to continually adapt the project delivery path. To reduce the risk of breakdowns in the information flow, approaches to streamline this flow are continually evolving. Current project engineering practice still relies on many tools that were developed in the pre-digital era, including project logs and scaled-down models of the project, such as scaled architectural models, scaled models of architectural details, and mock-ups. However, modern information and communication technology is changing how project information is handled. CAD systems allow designers to develop 2D or 3D models of a project's geometric features. BIM and CIM tools allow all project team members to integrate digitally created geometric models with other project documents and with information on the project context extracted from GIS or through laser scanning. Technology is also changing how traditional tools are produced. For instance, 3D printing is now allowing project teams to quickly go from digital to physical models to improve communications among team members.

Review questions

1. What is the main difference between 2D and 3D CAD?
2. What projects are a better fit for implementing BIM?
3. What projects are a better fit for implementing CIM?
4. Briefly explain the concept of AMG.
5. List the dimensions of a 4D model.
6. How can project owners mandate specific BIM-related contractual requirements?

16 Cost and schedule updates

Introduction

Construction managers use various terms in association with estimates and schedules, including "create," "report," "status," "update," and "revise." Some may see them as similar, but all are different, happening at different phases of design and construction and being performed by different individuals. In Chapters 17 and 18, we will discuss creation of the estimate and schedule, which are advanced project engineering functions. In Chapters 8 and 12, we discussed field and project engineers' roles with cost and schedule control, including how they assist superintendents in the field and report on the estimate and schedule to project managers and upper management throughout the construction process. In this chapter, we will discuss cost and schedule updates and particularly the role of the project engineer and some of the technology tools that enable him or her to perform these operations.

Cost updating

In an ideal world, if an estimator creates a perfect bid, the project is bought-out exactly according to that estimate. In such an ideal case, there are no change orders or deviations over the course of construction and the estimate might not require updating. However, construction is not based on exact science, which means there are not perfect bids and perfect design packages, so every project team is likely to see several updates to the original estimate and schedule during the course of the job. We have repeated Figure 6.2 here as Figure 16.1 for the convenience of the reader. We will refer to many of these steps or phases of the cost/schedule control process in this chapter.

Updating of the cost estimate starts almost immediately after submission of the original bid on a lump sum project or a cost proposal on a negotiated project. Errors are realized and must be corrected, and new information is discovered that can be incorporated into the estimate to make it even more accurate. In the case of a heavy-civil project, the estimator will do the same with respect to estimated individual unit price bid items but will also begin to keep a closer eye on the estimated compared to actual material quantities installed. An estimator develops a rough order of magnitude (ROM) estimate early in the bid process and continues to refine that estimate, by making it more accurate with the measurement of quantities and the incorporation of unit pricing and subcontractor bids. Similarly, the "final" estimate is not really "final" until the project is complete and closed out and the as-built estimate has been incorporated into the company's database.

After the contractor is notified that they are the successful bidder, they begin to carefully review and revise their original estimate to see if there are any overwhelming clerical or mathematical errors, which would allow them to not enter into a contract with the client. Depending on the jurisdiction, clerical or mathematical errors may allow a contractor to back out from a bid, whereas judgment errors usually do not. Once the contract is executed, the buyout team, which includes the project manager and senior-level project engineers, begins to interview subcontractors and suppliers. The buyout process will result in adjustments to many additional line items in the

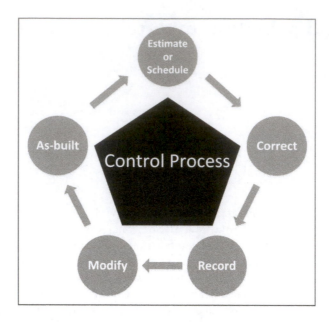

Figure 16.1 Cost/Schedule Control Process

estimate, some up and some down, but it will not change the contract value, especially in a bid scenario. The buyout and error corrections will be electronically incorporated into the bid estimate by the use of accounting tools, such as journal entries, with the estimate functioning as a balance sheet, where debits are matched with credits. As one cost code is increased by $1,000 (because it was lower in the original bid), two others must be reduced by $200 and $800, respectively, so the net impact to the bottom line is zero.

If a subcontractor, such as the painter, is bought out for $10,000 more than was included on the general contractor's (GC's) original bid (this can happen for a variety of reasons), then the contractor's cost is increased and the fee is reduced. This all occurs on the right-hand side of the following equation, but the contract value remains the same.

Contract Value = Construction Cost + Fee

This would be the case if there were only one change. But there will actually be several changes in both directions, some up and some down, all incorporated into the estimate with journal entries and all balancing out to zero. Estimators and experienced project managers are well aware that the home office is not interested in a reduced fee, rather just the opposite. These journal entries are included in the "correct" step in the cost/schedule control process.

Cost recording begins after the estimate is corrected and construction begins. As discussed in Chapter 8, cost recording involves:

- Cost-coding the direct-craft foremen timesheets,
- Cost-coding subcontractor invoices,
- Cost-coding material invoices, whether they be from long-form or short-form purchase orders, and
- Recording actual jobsite general conditions costs.

The cost process may be modified for a variety of reasons, but the contract value can change only when formal contract change orders are approved. Change orders are the topic of Chapter 20. Once a change order is approved, the contract value on the left-hand side of the equation will increase, if the change is adding to the construction cost. A corresponding increase is necessary on the right-hand side of the equation then as well, likely an increase in subcontract values, estimates for direct work, and the fee. The project engineer will likely not make these adjustments to the official estimate but rather will submit the necessary changes to the construction company's accounting department. This would be the "modify" step in the cost control process figure.

As discussed in Chapters 6 and 8 and other parts of this textbook, the development of the as-built estimate occurs as the last step of the cost control process and is a valuable tool to update the construction firm's database for the benefit of future estimates.

Schedule updating

The schedule update cycle follows a very similar path to the cost update cycle shown in Figure 16.1. The first schedule will be developed during the estimate or bid process to help establish the anticipated general conditions costs. Schedules are often required by clients with negotiated requests for proposals (RFPs) as well. Even contractors bidding unit prices for heavy-civil projects with owner-provided end dates and liquidated damages will rough out a schedule to determine the project's feasibility. All of these schedules are placeholders until the contract is awarded.

Once the contractor is notified of an intent to award by the project owner, a detailed contract schedule should then be collaboratively produced, as elaborated on in Chapter 18. Just as a construction team cannot perform cost control with an incorrect estimate, they must correct the schedule to make it easy to monitor and control, as will be discussed below. The steps necessary to prepare a detailed construction schedule include an analysis by the project superintendent, if he or she was not part of the original rough or summary schedule development at bid time. The superintendent will input his or her expected work plan and work flow and will perform manpower and crew size analysis. In addition, after subcontractor and supplier buyout, a collaborative approach to scheduling involves obtaining expected durations and material and equipment delivery dates from the GC's major second-tier team members. These dates should then be incorporated into the detailed project schedule.

In order for a schedule to be an effective construction management tool, it must be utilized and not just printed out and posted on the wall. This means it should be annotated weekly to monitor progress in relation to scheduled work. The superintendent customarily reports progress in the weekly owner–architect–contractor (OAC) meeting. Historically, a plumb-bob tied to a string has been used to show the advancement of the status line each week during the meeting. Colored highlighters can be used to visually indicate activities that are ahead or behind. Scheduling software programs also have means to depict schedule progress. The overall schedule status and any important issues should be documented in the weekly meeting notes, which were discussed in Chapter 11.

Schedules, like estimates, provide a map of where the project team wants to go and how they are going to get there. The objective is not necessarily to complete each activity exactly within its scheduled duration but to complete the overall project within the scheduled time. It is likely that many activities will be completed early and others late. Still, the jobsite project team wants to keep from finishing the entire project late; the objective is for the variances to average out.

Should the project team ever update or revise a schedule? If so, when and how? Responses to these questions depend upon both the contract requirements and the project progress. If the project is proceeding more or less on schedule, the schedule need not be updated. If the project is significantly behind schedule or there have been many change orders, the schedule should be modified and updated. Because computers have made it easy to revise the original contract schedule, it is now possible to compare the current schedule with the original contract schedule, which allows the project manager and project engineer to document schedule impacts when negotiating change orders. Sometimes it is necessary to revise the schedule, either because of a contract requirement or because of a substantial change in scope or progress. If it appears that construction progress is significantly behind, it may be necessary to develop a new schedule for a variety of reasons, including:

- Incorrect schedule logic,
- Inadequate estimate or schedule,
- Late material deliveries or lack of equipment availability,
- Client- or architect-induced delays,
- Additional scope and change orders, or
- Unusual weather, labor shortages, or inefficiencies.

Just as the project team should not knowingly work from an inaccurate cost estimate, they should not use an incorrect construction schedule. The construction team must first understand what the problems are and then endeavor to correct them. This may be done through a variety of methods, including:

- Revising the schedule logic,
- Paying extra for expedited material deliveries,
- Increasing manpower, working double shifts, and/or working overtime, or
- Removing and replacing a subcontractor, supplier, or craftsmen.

The project team must carefully assess the impacts of each method selected. Schedule recovery can usually occur only by accelerating or compressing activities on the critical path, as will be discussed in Chapter 18. Most of the above revision methods are not without increased costs and other impacts or risks, including production inefficiencies. The original schedule should be maintained intact if at all possible, for as long as possible. If a revised schedule is developed, it must be submitted to all subcontractors to be evaluated for impact on costs and means. Upon acceptance, the new schedule must be incorporated into every subcontractor and supplier's contract. Similarly to the incorporation of new drawings into contracts, this is a change order opportunity for subcontractors.

Home-office reporting

The jobsite management team is responsible for reporting to the contractor's home office on a variety of topics and to various levels of management; cost and schedule updates are at the forefront. In the case of an open-book negotiated project, many estimate and schedule updates and monthly status reports should be made to the client. The contractor's leadership responsibilities are also relevant to the corporate shareholders, investors, and bank and bonding companies. The project manager and project engineer input to the home office through journal entries, change orders, pay requests, and cost coding invoices, as discussed above. In addition, the project manager will submit a monthly forecast to revise all cost code projections and confirm the contract fee. All of these inputs will result in the home office developing a monthly job cost history report, which will be submitted back to the project team to confirm current accounts and statuses. The jobsite team will then do an audit of this report, which goes back to accounting, and the cost control cycle repeats.

Schedules should be maintained in a manner similar to the way that record drawings and as-built estimates are maintained. They should be marked up and commented upon to reflect when materials were delivered, activities started, activities finished, rain days, crew sizes, and whatever else the project manager and superintendent feel is appropriate. Heavy-civil unit price projects must also record actual material quantities installed. The as-built schedule, similarly to the daily job diary, becomes a legal tool in the case of a dispute. The recording of actual durations can also be used to more accurately develop future schedules and should be shared with the home-office scheduler. The project engineer can assist with all of these processes.

Project engineering applications

Similarly to many concepts in construction management, the ultimate responsibility of cost and schedule control, as well as subsequent updates, lies with the project manager on the office side and the superintendent on the field side. But project engineers often have technology skills in their toolboxes that their senior mentors might not have. In current practice, the use of technology—such as cost databases, accounting analysis tools, web-based reporting tools, and scheduling software—is prevalent and a project engineering responsibility. These skills enable them to greatly assist with updating the estimate and schedule. Some of the specific activities that the project engineer can help with include:

- Participating in the buyout process by attending meetings and recording changes that impact the overall estimate or schedule,
- Completing journal entries to submit to the accounting department,
- Cost coding actual subcontract and supplier expenditures and assisting with direct labor time sheet coding,
- Assisting with the monthly forecast through the development of field-foremen work packages and soliciting field input on remaining cost and time to go,
- Maintaining the client–GC and the GC–subcontractor and GC–supplier change order proposal and change order logs, and/or
- Contributing to the as-built estimate and as-built schedule development during the contract close-out phase.

Summary

Just as a broken hammer or dull saw are ineffective construction tools, so are inaccurate estimates and schedules. These documents need to be fluid to be effective construction management tools. The cost/schedule control process is circular and the starting point depends upon an individual's experience and the longevity of the construction company. The steps in this control process, as depicted in Figure 16.1, include (1) Estimate or Schedule, (2) Correct, (3) Record, (4) Modify, and (5) As-built, and then go back to (1) Estimate or Schedule.

Changes to the estimate and schedule begin shortly after the project is bid, when errors are corrected and buyout commences. The changes made to the estimate are through journal entries, where the additive changes have to equal the deductive changes, because the contract bid amount is firm. Change orders will revise the project estimate, but this will happen on both sides of the construction equation; essentially, the estimate functions as an accounting balance sheet.

Updating or revising the schedule can be an expensive task and does not need to be a monthly occurrence; if the project remains on track, leave the schedule be. Use of the schedule to monitor progress is the superintendent's responsibility and occurs during the weekly project owner–contractor coordination meeting, but the project engineer can assist with annotating the contract schedule for input to the as-built schedule during the close-out process.

Review questions

1. Who indicates progress on the schedule? When should it be done?
2. When should contractors avoid revising and reissuing the schedule?
3. When is it appropriate to revise the schedule?
4. How is an as-built schedule developed, and when should it be developed?
5. Explain how a journal entry will balance out, even though your elevator subcontract cost has increased by $20,000.
6. If the GC underruns one line item in the cost estimate, i.e., saves $5,000 on foundations, how does it affect the contract value for the following contract types? Assume all other costs and quantities remain as is and as estimated.
    a. Lump sum
    b. Cost plus with guaranteed maximum price, including an 80–20% savings clause (Note: 80% of savings would be retained by the project owner and 20% by the contractor)
    c. Cost plus with a 10% floating fee
    d. Unit price

Exercises

1. By utilizing our case study summary estimate for the Rose mixed-use apartment project from the companion website, prepare a revised estimate incorporating the following journal entries:
    a. Two storm water detention tanks are needed at four months each
    b. Structural steel was bought out at $2,100 per ton
    c. You have hired one subcontractor to do all glass, glazing, and related work for a lump sum of $800,000
    d. Our electrical subcontractor will take care of all temporary power and lighting requirements for a lump sum of $50,000, exclusive of monthly utility bills
2. Ignoring exercise one, incorporate the change order proposal from Chapter 20 into our case study contract estimate. What is your revised fee?
3. By utilizing the Rose mixed-use apartment project detailed schedule from the companion website, and without changing any logic and/or durations in the schedule, incorporate the following changes/updates and prepare a report to the home office regarding the project status. Are we ahead or behind schedule, and by how much?
    a. Elevator delivery is delayed one month, and they need two extra weeks for installation
    b. Exterior concrete walls were completed 11/1/18
    c. The fire protection subcontractor needs 18 days per floor because of concealed heads in the elevator lobbies
    d. The vinyl floor subcontractor has agreed to five more installers, which cuts their time for each floor by half
4. Utilize the schedule updates in exercise 3 and approach the following task similarly to an accounting journal entry; i.e., you need to maintain and/or report the original contract completion date without modification. What reasonable revisions to logic and/or durations do you make (a) if we are running behind schedule or, conversely, (b) if we are ahead of schedule?

Section D

Advanced project engineering

17   Cost estimating

Introduction

In addition to time, safety, and quality, cost is one of the four pillars to project success, and therefore, it is the focus of the project manager, superintendent, and project engineers. Cost control begins with the development of a detailed and accurate construction cost estimate. Project costs are estimated to develop a realistic budget within which the jobsite team must construct the project. All project costs are estimated in preparing bids for lump–sum or unit price contracts and negotiating the guaranteed maximum price (GMP) on cost plus contracts. Even when a GMP is not requested under a cost plus contract, the contractor is expected to provide cost forecasts to the project owner to facilitate their cash flow management.

Estimating can be summarized as the process of: counting and measuring material quantities from the drawings; applying competitive unit prices to those quantities; adding all the pricing up to a total, which is ideally a reasonable projection of the final cost of the project; and adding any markup to guarantee a profit. Cost estimates may be prepared either by the project manager and senior project engineers or by the estimating department of the construction firm. Even if the estimate is prepared by the contractor's estimating department, the jobsite team must understand how the estimate was prepared, because they will be tasked with building the project within the budget. This project budget becomes the basis for the cost control system that has been discussed throughout this book.

There are several good books, reference guides, software programs, and commercial databases available for construction estimating. Many contractors develop their own process and use customized spreadsheets that are supported by historical in-house databases. Recognizing how diverse each contractor's approach to estimating may be, we offer in this chapter just a brief introduction to some of the estimating activities that may be assigned to project engineers.

Because contractors operate for profit, it is essential for any construction management professional to acquire good cost estimating skills. The quantity take-off procedure for a few major building and civil assemblies will be briefly discussed in this chapter. If an estimator can calculate the quantity of concrete in a spot footing or count the quantity of hollow-core wood doors, then he or she has the skills needed to measure almost any material quantity. The same rules of measuring, counting, and pricing apply to most systems.

Although it may be difficult for some project owners to accept, there is not necessarily one correct total estimate, because numerous variables are subjectively dependent upon the estimator and the project. Therefore, some estimates are more correct than others, but estimating is not an exact science. Adjustments in pricing, subcontractor and labor strategies, overhead structures, and fee calculations are individual contractor decisions that will determine "the estimate" for those conditions at that time.

Estimating risk and strategy

Chapter 21 will provide an in-depth overview of how to manage project risks. Moreover, Chapter 8 identified the control of direct labor as one of the contractor's biggest risks and the estimating of the anticipated cost of that labor as one of the greatest challenges for the estimator. This is where an experienced estimator has an advantage. Moreover, estimating is often performed as part of competing for a contract, so the estimating strategy is paramount to a successful competition.

Other estimating risks involve failure to include some element of work or double-counting another element of work. To minimize errors when developing an estimate, the estimator should:

- Develop good estimating practices and procedures,
- Ask other experienced in-house project managers and superintendents to assist with the estimate,
- Request bids from only qualified subcontractors and suppliers, and
- Devise ways to build faster and beat the client's schedule, which will reduce the jobsite general conditions estimate.

Similarly to how the planning process precedes scheduling, there are many factors that will affect the estimating strategy for a specific project. Some of these factors include:

- The client, architect, and engineers,
- Contract format,
- Project location,
- Complexity and accuracy of the drawings and specifications, and
- Quantity of construction projects under contract but not yet complete.

Types of cost estimates

The project manager who will be assigned to build the project, along with his or her project engineers and superintendent, should be responsible for developing or providing input to the estimate. Their individual contributions regarding constructability and their personal commitments to the estimating product are essential to assure not only the accuracy of the estimate but also the ultimate success of the project. There are several different types of cost estimates. The three most common estimates are:

- *Conceptual cost estimates*: These are often developed several times during design development by using incomplete or preliminary drawings and specifications, such as schematic design documents. They are used to set a project budget and verify its accuracy throughout early design stages. Because they are only indicative of the final cost, they can be used to set a starting budget on cost plus contracts that are awarded based on qualifications;
- *Detailed cost estimates*: These are prepared by using complete drawings and specifications, or construction documents (CDs), and are used to respond to bids for lump-sum and heavy-civil unit price contracts; and
- *Semi-detailed estimates*: These include elements of both conceptual and detailed estimates, and they are used to provide a GMP on a cost plus contract or a lump sum price for heavy-civil design–build contracts. They are usually developed to participate in best value procurement.

The procurement approach and the contract type may seem to affect the type of estimate to adopt in any given case, but it is really the level of design detail that indirectly affects this decision, as well as the procurement and contract. As a matter of fact, it would be impossible to perform a detailed cost estimate for a conceptually designed project. However, it is possible to do just the opposite—i.e., develop a conceptual estimate out of a nearly complete design—and it is actually an early step of the detailed estimating process that will be discussed later.

The estimating strategy is different with each of these estimate types. The level of detail will correspondingly differ with each of the estimate types as well. Early estimates based upon preliminary documents include many uncertainties in the form of *allowances*, *contingencies*, square footage pricing, and assembly estimates. The more advanced and detailed the drawings, the more the estimator will utilize historical labor productivity, wage rates, market material pricing, and competitive subcontractor quotations.

The accuracy of an estimate is also directly proportional to the quality of the documents. The time spent on preparing an estimate also follows those same lines. Conceptual estimates may, at best, provide an accuracy within ± 20% of the final construction cost; semi-detailed estimates are, at best, within ± 5–10% of accuracy; and contractors expect to be within only a few percentage points of the final cost with their detailed estimates.

Every estimate has similar major elements or categories, including heavy-civil unit price bids. These categories include:

- Direct labor,
- Direct material,
- Direct equipment,
- Subcontractors and major material suppliers,
- Jobsite administration or general conditions, including equipment that cannot be allocated to individual tasks, and
- Percentage markups, including the fee, taxes, contingency, and insurance.

Conceptual estimates

To those individuals who are not familiar with how the construction industry operates, all contractors' estimates are viewed the same, in that they are all "bids" and all contractor-produced estimates are detailed and completely accurate. This of course is a misconception because estimates are dependent upon various factors, including the quality and detail of the design documents. Early conceptual estimates may be developed by contractors or architects on limited information and produced with little effort. These estimates are not expected to be "firm" or very accurate. Instead, they should be used to guide conceptual and schematic design to stay within the approved budget, rather than as firm bids. If subsequent conceptual estimates show that the initial budget cannot be met, they can be used to go back to the owner to request either an upward revision of the budget or a downsizing of the project scope. A conceptual set of drawings can be quickly estimated as follows:

- Early on, historical unit prices for square feet of floor or for cubic feet of building volume can be used to develop an early estimate.
- When design has advanced enough to allow for a work breakdown among specialties, values can be used as placeholders for each specialty, also referred to as *subcontractor plugs* or, sometimes, *subcontractor allowances*; these values are only educated guesses for the project owner,

who would need to reevaluate once design progressed. Ideally, these values will be replaced by subcontractor bids at the end of the detailed estimating process.
• Percentage add-ons for jobsite administration and markups are used throughout this process.

Most items within a conceptual estimate are usually allowances. Conceptual estimates are correspondingly the least accurate estimate type and should carry substantial contingencies. They are developed by seasoned estimators, whereas junior estimators may perform quantity take-offs (QTOs) to support detailed estimating.

Even when a semi-detailed or detailed set of drawings is provided, it is customary for a contractor to also develop a conceptual estimate as soon as the design package is available. Although this estimate is not used to submit a final price for the project, it is utilized to determine if the project is the right size for the contractor to pursue and which of their estimating or management staff would be best suited to work on the detailed portions of the estimating process.

Detailed estimates

Much of the focus of the remainder of this chapter will be on the process of producing detailed estimates. A detailed estimate takes the longest time to prepare, costs the contractor the most in personnel resources to produce, and produces the most accurate final figure. Drawings and specifications that are at least 90% complete are ready for a detailed estimate. A fully detailed set of drawings is needed to prepare a detailed estimate so that every item of work can be accurately measured and matched with unit prices or competitive subcontractor bids. Detailed estimates are most prevalent on projects that are bid either as a single lump sum price, for a new high school as an example, or as a set of unit prices for heavy-civil projects, such as a highway overpass.

Lump sum bids

Lump sum bids are also referred to as "hard bids," "fixed price," or "stipulated sum" and are typically associated with certain types of public works projects. In a slow economy, project owners on privately funded projects may solicit lump sum bids and still secure a quality general contractor. In a busy economy, however, they can secure a quality contractor only through negotiated requests for proposals (RFPs). Project owners requesting a lump sum bid are primarily interested in just the bottom line and assume that all contractors can deliver the project with comparable levels of quality, schedule, and safety. As shown in Figure 17.1, the detailed lump sum bid estimating process includes:

• Quantity take offs (QTO),
• Direct work pricing recaps,
• Subcontractor plugs and later bids,
• Jobsite indirect general conditions, and
• Markups.

Unit price bids

Unit price bids follow a similar process to lump sum bids, except individual standalone bids are created for each line in the estimate. Unit price bids typically apply to public works and utility projects where the drawings will only allow an estimator to compute quantities within a range of uncertainty. Before letting the project go to bid, the client employs an estimating consultant, or quantity surveyor, to develop quantities for each line item that all of the contractors are to utilize. By taking the risk of measuring quantities away from the contractors, this approach subsequently

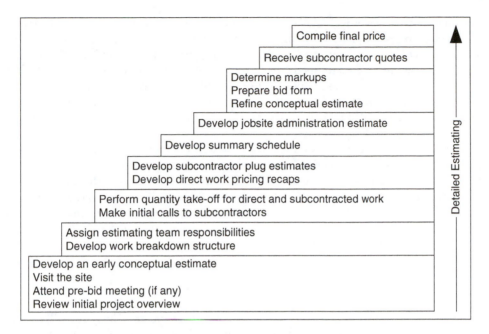

Figure 17.1 Estimating Process

lowers their risk and, theoretically, their contingencies. Projects prone to utilize the unit price method are typically linear heavy-civil projects, such as projects for highway pavement overlays or installation of storm drains and sanitary sewers. It is not unusual to see a bid form with 100–200 separate line items that the contractor must develop standalone pricing for. The contractor develops lump sum estimates for each of these line items, including direct labor, material, equipment, and subcontractor pricing. Typical markups, including the fee and insurance, are also added to each line item, and the totals are extended across the spreadsheet to the right, to create a comprehensive unit price. These complete standalone unit prices are then applied to the client's quantities. All line item bids are then totaled for one complete lump sum bid. If the actual quantities installed during construction vary from the bid quantities on any single item, then the contract value is adjusted by utilizing these individual prorated unit prices. This is the type of estimate developed by Gateway Construction for the I-90 overpass case study project described throughout this book.

Semi-detailed estimates

Semi-detailed estimates are a hybrid of conceptual and detailed estimates. They are used under various delivery methods and contract types. For instance, they are typically used to provide a GMP on a cost plus contract. They are also used to provide a lump sum price for heavy-civil design–build contracts. By utilizing measured quantities and unit prices, they are produced to price work packages that have been adequately designed and specified, such as structural concrete. Subcontractor bids are utilized wherever possible, but subcontractor allowances are included for areas not yet completely designed, such as interior finishes. For work that is not fully designed, the contractor will use assembly costs, such as square footage estimates, to verify allowances.

Reliance on contingencies is more prevalent in semi-detailed estimates than in detailed estimates. When used to produce a GMP for a cost plus contract, the contingencies are known by the project owner and the contract is worded to incentivize the contractor to reduce withdrawals from contingency accounts by providing a share of any savings, such as 80% to the project owner and 20% to the general contractor. In these contracts, a definition of costs that are reimbursable and those that are non-reimbursable must be established in the contract. Also, costs that are considered jobsite general conditions versus those that are considered home office and part of the fee are paramount and should also be established in the contract before work commences. A GMP was negotiated between the developer and Reliable Construction for the Rose Collective mixed-use apartment case study discussed in this and other chapters.

Semi-detailed estimates are also used by design-builders competing for a lump-sum contract for heavy-civil public works. These contracts are usually awarded on a schematic design and require prospective design-builders to advance the design for submission jointly with a price proposal for a lump sum. Given that it would be uneconomical to fully develop the design in the absence of a contract, prospective design-builders will also rely on contingencies and allowances. Contingencies and allowances, however, will not be shared with the project owner, who is accepting the potential extra cost in exchange for an early price certainty.

Process

The process of developing a detailed estimate is shown in its entirety in Figure 17.1 and can be followed when the design is ready for a detailed estimate. However, the same process can be used to develop semi-detailed estimates by skipping those steps that cannot be performed due to lack of information.

The initial steps of gathering information provide a foundation for the whole process. One of the first assignments for the estimating team is to develop a responsibility list and to schedule the estimate. The estimating process should be scheduled for each project beginning with the dates for the pre-bid or pre-proposal conference and the date that the bid or proposal is due. With these milestones established, a short bar-chart schedule can be developed that shows each step and assigns due dates to the estimating tasks depicted in Figure 17.1. Familiarity with the steps or building blocks is essential in developing the schedule. Each team member is relying on the others to do their jobs efficiently, accurately, and in a timely manner. Similarly to a construction schedule, if one of the individuals falls behind on any one activity, the completion date may be in jeopardy or, more commonly, the quality of the finished estimate will be affected, unless other resources are applied.

As the estimating team proceeds through the process, information continues to be analyzed, summarized, and refined, until eventually there is only the final estimate or bid. Here we will elaborate on some of the most crucial steps that lead to this outcome.

Work breakdown structure

The work breakdown structure (WBS) can take various forms. Graphically, it can be represented as "a hierarchical decomposition of the total scope of work to be carried out by the project team to accomplish the project objectives" (PMI 2016; p. 14). However, it may more simply take the form of a listing of all the work activities shown on the drawings and in the specifications. It is often prepared in outline form with variable "levels" or degrees of detail, depending on the mix of self-performed versus subcontracted work and the completeness of the documents. It includes areas of work, such as foundations, utilities, drywall, floor covering, and plumbing. Before any detailed estimating is performed, including the quantity take-off, the estimator should have a general idea of all the work that will be included on the WBS. The first step is to conduct a complete document overview. The goal is to review the drawings and specifications thoroughly and develop a good

Project: I-90 Overpass, Project #1732   Date: 3/25/2018
Estimator: Jim Bradley

Line #	CSI Div	Cost Item Description	General Contractor Materials	General Contractor Labor	Subcontractor Materials	Subcontractor Labor
2	2	Sitework	X	X	X	X
3	3	Concrete:				
3.1	3	CIP Concrete:				
3.1.1	3	Pile Caps:				
3.1.1.1	3	Excavate	X	X	X	X
3.1.1.2	3	Formwork	X	X		
3.1.1.3	3	Rebar	X	X		
3.1.1.4	3	Anchor Bolts	X	X		
3.1.1.5	3	Purchase Concrete	X			
3.1.1.6	3	Pump Concrete			X	X
3.1.1.7	3	Place Concrete		X		
3.1.2	3	SOG	X	X	X	X
3.1.3	3	Columns	X	X		
3.1.4	3	Elevated Decks	X			X
3.1.5	3	Gypcrete			X	X
3.2	3	Pre-cast Concrete		X	X	
4	4	Masonry			X	X
5	5	Structural Steel:				
5.1	5	Rolled Shapes:				
5.1.1	5	Fabricate	X			
5.1.2	5	Install				X
5.2	5	Metal Deck	X			X

Figure 17.2 Project Item List

understanding of the type of project and the systems that are included. The estimator should not start quantifying or pricing any work items until this overview is complete.

Many estimators format a WBS in outline form, but a project item list is one alternative way to present the WBS. Figure 17.2 is an abbreviated example for our heavy-civil case study project. This list is a good reference to use throughout the estimating process and again as a final checklist to review prior to finalizing the estimate. This item list is not to be considered as a final WBS, just a good first step. The WBS will continue to evolve throughout the estimating, scheduling, and buyout processes. As the estimator dives more deeply into the project, there will be several more detailed levels of the WBS. A good WBS or project item list can also be utilized later to develop the project schedule, as will be discussed in Chapter 18.

At this stage, the estimator needs to know which categories of work will be self-performed by the contractor's own forces and which will be supplied and installed by subcontractors. The project item list has boxes to check for this purpose. Categories of work that general contractors may perform on building projects include:

- Minor structural hand excavation and backfill,
- Concrete formwork,
- Concrete reinforcement steel placement,
- Concrete placement,
- Concrete slab finishing,
- Structural steel erection,
- Rough carpentry, and/or
- Finish carpentry, including doors, specialties, and accessories.

Heavy-civil contractors may also perform earthwork, shoring, paving, and utility work with their own direct forces, because they may have access to specialized equipment for those operations. The estimator and the construction team decide which work items should be self-performed or subcontracted by evaluating several issues, including:

- *Labor*: Subcontractors must be used for a work item if the tradesmen needed are not employed by the general contractor. Although shifting more work to subcontractors is a means of risk mitigation, there are also additional risks when subcontracting, which are discussed in Chapters 19 and 21;
- *Specialization*: If a subcontractor specializes on a work item, such as finishing concrete, they may have refined the most efficient and least expensive way to perform the work;
- *Quality*: If this is a work item that the subcontractor specializes in, they can often also perform it better. Also, if there are problems with quality, a subcontractor is required to repair the work without increase in cost to the general contractor. Conversely, if the general contractor is required to repair self-performed work, such as a weld that failed inspection, the general contractor's cost will increase;
- *Price*: If a subcontractor can perform the work on a fixed price contract for less than what the general contractor has estimated, then the work may be awarded to the subcontractor;
- *Work load*: If the general contractor's labor forces are tied up on other work, some normally self-performed items may be awarded to subcontractors. Conversely, if the general contractor's volume is low but they want to keep the labor force together, the GC may choose to self-perform more work items; and
- *Safety and schedule*: Many general contractors will argue that they can control and assure project schedule and safety better with their own forces.

Once the mix of direct versus subcontracted work has been determined, select subcontractors will be sent solicitations to prepare bids; this is often a role for an experienced project engineer. It is a good idea to have the subcontractors working to prepare their estimates in parallel with the general contractor developing their own estimate for self-performed work. When subcontractors are notified, they will ask the estimator questions regarding specifications, quantities, and materials. The project engineer or estimator should be informed about specific subcontractor scope, but each subcontractor should develop a completely independent estimate. Incorporation of the cost of subcontracted portions of the work into the project estimate is discussed in detail later in the chapter.

Quantity take-off

After the WBS has been completed, the estimator proceeds to count and measure material quantities from the drawings. The goal is to perform detailed measurements and counts for each item of work that has previously been selected to be self-performed and to record them on quantity take-off (QTO) sheets. The QTO process is one of the most time-consuming activities in the detailed estimating process shown in Figure 17.1 and is a necessary early step toward preparing the final total estimate value.

The proper order of preparing the QTO is customarily the order in which the work will be performed, such as the foundations before the slab on grade (SOG) before the columns. This is typical for commercial, civil, or residential projects. This order will accomplish several tasks, including forcing the estimator to think like the superintendent, that is, the floor system is built before the walls. Organization of the estimate in this fashion will later assist with the schedule development and will aid with the development of the project cost and schedule control systems. All work items should be taken off prior to pricing. Material quantities are recorded and later extended out and summarized on QTO sheets similar to the one shown in Figure 17.3. Good QTO procedures include:

- The drawings should be marked up, whether done by hand with colored pencils and highlighters or shown on the computer, to indicate what items of work have been taken off. This will help minimize errors;
- Sketches and assumptions should be developed and preferably noted on the QTO sheets;
- Quantities should be measured, extended, and summarized as they will be priced and purchased. For example, concrete is initially measured in cubic feet (CF), but it is purchased in cubic yards (CY). Concrete reinforcement steel is quantified by the piece, multiplied by the length per piece, multiplied by the weight in pounds per linear foot, and finally converted into tonnage. Waste and/or lap are added if appropriate;
- The estimate should appear neat and professionally organized, because upper management will want to review the estimator's work prior to bid for completeness and accuracy; and
- Although most estimating activities are being performed electronically today, paper print-outs are inexpensive and are a useful method to perform a quality-control check of the estimate.

Every estimator has his or her own "rules-of-thumb" with respect to the amount of waste or lap to apply to measured material counts. An allowance of 5–10% is common and should be included on the QTO sheets. Purchasing enough, but not too much, material is important to maintain labor productivity. Items such as nails, glue, caulking, and rebar tie-wire can be estimated or allowed for, but determination of exact quantities is difficult and the amount of time the estimator would need to estimate them would usually exceed the value of the materials. Allowances are usually sufficient at this stage.

The QTOs developed by the GC for subcontracted work are not as detailed as those for its own direct work. On lump sum projects, GCs develop early in-house estimates for subcontracted work that will be used initially as plugs but will later be replaced with competitive quotes when they are received.

Pricing self-performed work

Each contractor will have their own standard pricing recapitulation sheets—also known as recaps—to extend these quantities and apply labor and material unit prices. An example of a pricing recap sheet for our heavy-civil case study is shown in Figure 17.4. All the recapped material quantities

Gateway Construction Company
Quantity Takeoff Sheet

Project: I90 Overpass   Date: 3/25/2018
Owner/Loc: State of Montana, Missoula, MT   Estimator: Jim Bradley
                                                       Est #:   1

Div #:   03, Pile Caps, Bid Item 172

Ref	Description	Qty (ea)	L (ft)	W (ft)	H (ft)			
	15 each: 2.5'd x 10' SQ with #6 @ 6" OC rebar EW T&B and 8 EA 1" round x 12" long ABs/pile cap:							
	Structural Excavation: Top of pile cap 2 feet below existing rough grade							
	Pile Cap Size:	15	10	10	2.5			
	Hole Size:	15	12	12	4.5	9720	CF	
		Plus 40% swell @ 27CF/CY=				504	Say 500 TCY	
						SF	CF	
						Formwork:	Concrete	Finegrade:
	Concrete:	15	10	10	2.5	1500	3750	1500 SF
								@ 9SF/SY=
								167 SY
	Concrete purchase:	@ 1.055% waste @ 27CF/CY =					147 CY	
	Rebar:	4 matts @ 21 bars per matt @ 10' long per bar =						840 LF/PC
		@ 15 pilecaps =						12600 #
		#6 @ 1.502#/LF @ 2000 #/T =						9.5 tons
		Note: no lap or waste figured with spot footings or pile caps						
	Anchor Bolts:	8 per PC =		120 EA				

Figure 17.3 Quantity Take-Off Sheet

are brought forward from the QTOs and entered onto the pricing sheets. The estimator should not start pricing until all the materials have been taken off the quantity sheets and entered on the pricing sheets. Similarly to the process for taking the materials off the drawings, each quantity is circled or highlighted to indicate it has been brought forward to a pricing recap sheet. After all of the QTO sheets have been brought forward, the estimator begins labor and material pricing.

An alternative ten-page detailed estimate format for our Rose Collective mixed-use development (MXD) apartment case study is included on the companion website. In this example, the separate labor and material unit price columns have been collapsed into one assembly unit price for each of the several hundred line item estimate entries.

As discussed, the contractor's direct craft labor productivity is the most difficult item to estimate and control, and it is therefore the riskiest item. Often a general contractor will review the amount of labor in an estimate and use this figure as some basis for determining the overall project risk, thereby determining the appropriate fee to apply. Labor productivity is estimated as man-hours/

Gateway Construction Company

Project: Highway 90 Overpass
Location: Missoula, MT
Engineer: State of Montana
Division/Category: 03: Pile Caps

Bid Item: 172

Date: 3/25/2018
Estimator: Jim Bradley
Estimate #: 1

15 each: 2.5'd x 10' SQ with #6 @ 6" OC rebar EW T&B and 8 EA 1" round x 12" long ABs/pile cap:

Description	Qty	Unit	Material Rate	Material Cost	Labor UMH	Man Hours	Wage Rate	Labor Cost	Equip Rate	Equip Cost	Total Cost
Layout	1	LS	250	250	16	16	41	656			$906
Structural Excavation	500	TCY		0	0.085	43	40	1700	7.5	3750	$5,450
Hand Ex/Fine Grade	167	SY		0	0.05	8	33	276			$276
Forms	1500	SFCA	1.25	1875	0.075	113	41	4613			$6,488
Rebar	9.5	Tons	1200	11400	15	143	50	7125			$18,525
Anchor Bolts	120	EA	18.1	2172	0.25	30	41	1230			$3,402
Purchase Concrete ★	147	CY	95	13965		0		0			$13,965
Pump Concrete	147	CY		0		0		0	12.5	1838	$1,838
Admixtures	147	CY	2.5	368		0		0			$368
Place Concrete	140	CY		0	0.33	46	33	1525			$1,525
Protect and Cure	1	LS	300	300	5	5	33	165			$465
Strip Forms	w/forms										$0
Backfill	353	TCY		0	0.1	35		0	6.5	2295	$2,295
				30330		438		17289		7882	$55,500

Wage Check:
Labor $ / Hours =   39.44 $/HR. Checks? **Yes**

System Check:
Total $ / CY =   396 $/CY   Checks? **Yes**

★ Although the stated quantity of 140 cubic yards is reflected on the bid form SOV, we have to allow an additional 5% for waste in our workup calcs

Figure 17.4 Pricing Recapitulation Sheet

unit of work, such as six man-hours (MH) per door. This is referred to as unit man-hours (UMH). This system allows for fluctuation of labor rates, union versus open shop, geographic variations, and time. If it takes 3.25 UMH per door to install an interior hollow metal doorframe in Montana, it probably takes the same in Mississippi. Appropriate wage rates can be applied for the specific project. The best source of labor productivity is the estimator's in-house database. Each estimator or contractor should develop their own database from previous estimates and previously developed labor factors. Other sources of UMH data would include published databases or reference guides, such as RS Means' *Building Construction Cost Data.*

Note: As we mentioned before, in an effort to use gender-neutral language to reflect an increasingly diverse workforce, some companies are replacing the term man-hour with person-hour (PH). This terminology is less common and not widespread, so we decided to follow the traditional terminology in the rest of this book.

Extended man-hours should be rounded off. Fractional man-hours should not be retained on the pricing recap sheet. Fractional man-hours are difficult to schedule and monitor against for cost control, let alone difficult to explain to the superintendent why he or she has 1.7 man-hours to place the gravel for the perforated drain system and not 2 man-hours. As already stated, estimating is not an exact science; although it relies on sound math, good judgment and procedures are equally important. The man-hours are totaled at the bottom of each pricing page. This figure will be valuable information later on for scheduling and cost control.

Every contractor will have their own direct wage rates. These vary according to location, union agreements (if applicable), and the type of work to be performed. Sources of wage rates may include prevailing wage rates, union rates, merit shop rates, or in-house rates. On non-prevailing wage projects, only bare (paid to the employee) wage rates are included on the pricing sheets. Labor burdens or labor taxes are included on the estimate summary page. Labor burden will add 30–60%, depending upon craft, on top of the labor estimate for items such as workers' compensation, union benefits, unemployment, Federal Insurance Contributions Act (FICA; social security), and medical insurance. These percentages fluctuate with time and location. The estimator should check with his or her company's current accounting data before applying a percentage. Prevailing wage projects have labor taxes (but not labor benefits) built into the rates. The prevailing wage rates for our Montana overpass case study project, which were in place in 2017, are shown in Table 17.1.

Competitive market-rate material unit prices should be solicited from local and specified suppliers. Suppliers are requested to provide prices for substantial materials that will be purchased by the general contractor. Price quotations should be received in writing on the supplier's letterhead. If telephone quotations are received, company-consistent forms should be used for recording material pricing, although pricing on the vendor's letterhead is always preferred. Many material

Table 17.1 Prevailing Wage Rates

Craft	Base Wage	Fringe Benefits	Prevailing Wage Rate
Carpenters	$29	$12	$41
Cement Masons	$22	$11	$33
Equip. Operators	$28	$12	$40
Laborers	$24	$9	$33
Electricians	$33	$13	$46
Ironworkers	$27	$23	$50
Plumbers	$36	$17	$53

quotes are received electronically. Other sources of material prices include in-house historical costs or previous estimates. Historical prices are not as reliable as current quotes, but it is better than leaving an item blank. The third choice for material prices are databases or reference guides. These sources are useful for unusual items for which local quotes cannot be obtained, as allowances or as subcontractor plugs in early budget development.

Figure 17.4 shows a pricing recap sheet for the I-90 pile caps. Each of the quantity line items is extended across. For example: A measured material quantity of 1,500 SFCA (square feet of contact area) pile cap formwork has been carried forward from the QTO to the pricing recap page. The material prices are extended by multiplying this quantity of formwork by the material supplier's quote of $1.25 per SF to yield a material price of $1,875. The measured quantity is also multiplied by the historical UMH of 0.075 MH/SFCA to yield 113 man-hours. The carpenter's prevailing wage rate of $41 per hour is applied, which yields a total of $4,613 to form the pile caps. Fractional man-hours and cents are applicable in the unit price or UMH columns of the pricing recap page only, not the extended columns. Extended prices should be rounded to the nearest dollar. After all of the unit prices have been applied to the quantities, total costs are recapped for each page and/or each assembly or system, in preparation to be brought forward to the estimate summary page.

Pricing subcontracted work

General contractors rely on specialty contractors to perform a majority of the work as subcontractors, and the best source of pricing is from these specialists. Subcontractors are the ones who will ultimately be required to sign a contract and guarantee performance for a fixed amount of compensation. General contractors should estimate subcontractor work to develop pre-bid day budgets and to check the reasonableness of subcontractor bids. This is why general contractors also develop rough QTOs for subcontracted work. In-house GC estimates for subcontracted work are referred to as *subcontractor plugs* or simply *plugs*. If one low bid of $10,000 for flagging comes in and one higher bid of $20,000 is available, the general contractor may feel comfortable throwing the high bid out because their own pre-bid plug was $12,500. The reverse may be true on another system, if the general contractor's estimate was nearer the high bid. Once a reliable subcontractor price is received, the GC's plug estimate should be replaced with the subcontractor's price. Another situation would occur if the general contractor plug was significantly higher or lower than any subcontractor bid. In these cases, the estimator should not assume that the subcontractors, who specialize in painting, for instance, are all in error and that his or her pre-bid plug is the most correct one. A spreadsheet or bid-tab should be developed that aids in the analysis of subcontractor and supplier pricing.

Jobsite administration or general conditions

There are many different uses of the term general conditions, including general conditions to the contract and home-office general conditions. Jobsite general conditions costs are project-specific costs and are measurable. It was discussed earlier that estimating direct labor costs is one of the most difficult and riskiest tasks for the construction estimating team, but estimating jobsite general conditions is also difficult. Site-specific general conditions are also referred to by some as jobsite administration costs, indirect costs, distributable costs, or overhead costs. Jobsite general conditions are indirect costs that are necessary to manage and supervise the project but are not necessarily attributable to any specific area of work.

Even if the general contractor is not required to turn in a construction schedule, it is still necessary to develop a relatively accurate schedule prior to preparing a general conditions estimate. By utilizing past experience, coupled with the estimated man-hours and with the superintendent's input, it is a fairly simple process to prepare a 20–30 line item schedule. The project team will rough out a schedule based upon these hours, the project's complexity, and subcontractor input on durations and deliveries. This schedule is used to develop overall durations for estimating salaried administration labor, such as supervision and project management, as well as equipment rental (e.g., a pickup truck). It is not necessary to determine if it will take exactly 440 work days to build the apartment building, but it is important to know that this structure will take approximately 18 months to construct (not ten or 30) given the site conditions, project complexity, long-lead material deliveries, anticipated weather, and available manpower. A summary schedule for the Rose Collective MXD is presented in Chapter 18.

A collapsed summary of Gateway's I-90 jobsite general condition's estimate is shown in Figure 17.5. The complete general conditions estimate for this project is four pages long and is available on the companion website. That estimate version includes a generic comprehensive list of items, which may be considered jobsite general conditions. The complete MXD case study general conditions estimate is also included on the website. There is not an exact rule for what the total general conditions costs should be. They could range anywhere from 5% for a larger building project to 10% for a smaller building project, for an equipment- and labor-intensive civil or residential project, or for out-of-town projects. The distinction of which items will be jobsite reimbursable and which are part of the fee will be spelled out for a cost plus contract and will affect this range. Many of the individual general conditions line items are project- and site-specific and others are time-dependent.

Estimate summary

There are several final steps in completing the estimate. For building projects, a one- or two-page estimate summary will be utilized to gather all the direct work estimates, general conditions, subcontractor pricing, and markups. In the case of a 100–200 line item heavy-civil unit price bid, this final summary, and/or the bid form, may be several pages long.

An estimate summary form, similar to other estimating tools, should be developed for each project type and used consistently throughout the construction firm. It may require slight modifications for any one project, but its consistency is important to provide the project estimating team with a quick and comfortable overview prior to and during bid day. If a certain project has specific needs that the estimate summary does not account for, a project-specific summary must be prepared.

The estimate summary is often roughed-out at least one day prior to bid day. A pre-bid-day estimate summary is developed as a refinement of the first rough order of magnitude (ROM) conceptual estimate from the day the bid documents first came in. As the estimator completes the estimate, he or she should use the most current and relevant information. All the pricing recap pages are brought forward by posting values on the summary sheet. A hard copy should be printed out and saved.

At the bottom of the estimate summary page, below the subtotal of the direct costs, is a series of percentage markups including the fee. The fee is comprised of home-office overhead and profit (OH&P). Home-office overhead costs include items such as accounting, marketing, and officer salaries. These costs are usually relatively fixed figures based upon staffing for one fiscal year. An average size general contractor in the building sector may need to generate a minimum fee of 1.5–3% of their annual volume to cover their home-office overhead. In addition to covering the indirect costs, the contractor desires to earn a profit on each project. The desired profit plus the home-office overhead indirect costs is the OH&P fee. In the case of Reliable Construction

Gateway Construction
Jobsite General Conditions Estimate
(Abbreviated)

Project: I 90 Overpass, Project #1732
Owner: State of Montana

Estimator: Jim Bradley
Estimate #   1
Date: 25-03-18

Description	QTY	Units	Labor Unit Price	Labor Cost	Material Unit Price	Material Cost	Total Cost
From Page 1, Administration Expenses:							
Project Manager	60	wks	2100	126000		0	126000
Project Superintendent	60	wks	2200	132000		0	132000
Project Engineer	60	wks	1250	75000		0	75000
Safety Engineer	12	wks	1200	14400	1400	16800	31200
Field Engineer	See Project Engineer			0		0	0
Form Detailing	1	LS		0	2500	2500	2500
Office Utilities	14	mos		0	500	7000	7000
From Page 2, Equipment:							
Pickup Truck	14	mos			800	11200	11200
Compressor	w/cost of the work					0	0
Welder	w/cost of the work					0	0
Forklift	14	mos	1125	15750	3275	45850	61600
Small Tools	3,301,005	Labor $			1%	33010	33010
Consumables	14	mo			500	7000	7000
From Page 3, Temporary Construction:							
Job Office	14	un/mo			2000	28000	28000
Dry Shacks	14	un/mo			1800	25200	25200
Temporary Lighting	Separate stand alone bid item on SOV				0	0	0
Radios, 2 ea	28	un/mos			75	2100	2100
75kw Generator for Crew Shack	14	mos			2750	38500	38500
5kw Generator for tools, 5 each	70	un/mos			450	31500	31500
Fences	1950	LF/mo			7	13650	13650
From Page 4, General Operations:							
Street use Permit	14	mos			1225	17150	17150
Flaggers	Separate stand alone bid item on SOV				0	0	0
Periodic Cleanup	14	mos	1000	14000			14000

	Labor	Material	Total
TOTALS FROM PAGE 1, ADMIN:	347900	41400	389300
TOTALS FROM PAGE 2, EQUIP:	15750	97700	113450
TOTALS FROM PAGE 3, TEMP:	0	142950	142950
TOTALS FROM PAGE 4, OPERNS:	19400	79463	98863
GRAND TOTAL JOBSITE GENERAL CONDITIONS:	$383,050	$361,513	$744,563
LABOR BURDEN ON SALARIED LABOR:		$75,564	$75,564
ADDITIONAL BURDEN ON UNION LABOR:		$91,933	$91,933
TOTAL JOBSITE GENERAL CONDITIONS:			$912,060
% OF TOTAL BID @ $9,510,000 CHECKS? YES			9.6%

Figure 17.5 General Conditions Estimate

on our MXD project, a combined fee of 5% is included. Fees can range anywhere from 3% to 8% for commercial work and often higher for residential projects because of the increased labor risk. Smaller commercial projects require higher fees as a result of the smaller annual volumes of the firms involved. If the general contractor self-performs the majority of the work and/or utilizes their own equipment, as is the case with many heavy-civil contractors, they also have a proportional increased risk and fee. Gateway Construction included a 13% fee on the overpass project. Some of the other markups or add-ons that are applicable on many projects, be they bid or negotiated, include:

- *Material or sales tax*: This varies between states. Sometimes these values are already included in the pricing, or sometimes they apply only to materials that are not incorporated into the project (e.g., concrete forms);

- *Business tax or excise tax*: This is also job-cost related and is dependent upon the city, county, and state;
- *Insurance*: General liability insurance is volume-related. The insurance rate will vary significantly between contractors depending upon their size and safety records. This markup generally ranges from less than 1% for large commercial and civil contractors to 2% for smaller GCs and specialty contractors;
- *Contingency*: If the project is competitively bid with relatively complete documents, the amount of contingency applied is usually zero. Most general contractors account for the contingency, if any, with their choice of fee on competitive projects. Stated contingencies will show up on negotiated projects, which have incomplete documents; and
- *Bond*: Performance and payment bond rates are also annual volume- and company performance-related, but the project bond costs are project-specific. On larger lump sum projects, they may be required, especially for public and civil work, and the cost will range depending upon the size of the project and the past performance of the contractor. The bond cost is customarily "below-the-line"—that is, below the base bid—or an alternate add-on and not included in the bid. Bonds are frequently factored into individual unit price bids for heavy-civil jobs. Performance and payment bonds generally do not appear on private negotiated projects.

The final and total estimate value is determined on bid day or, in the case of a negotiated proposal, the day the proposal is due. This is the final step in Figure 17.1. Subcontractor quotations will be received, plugs will be removed, and the totals revised. The final bid is generally approved by an executive of the construction firm. The total figure must be submitted on the form specified in the instructions to bidders (ITBs) and submitted to the client before the specified time. An abbreviated summary estimate for our heavy-civil case study is shown in Figure 17.6, and its commercial counterpart is shown on the companion website. In a lump sum project, the summary estimate might be only one page, whereas a completely open-book negotiated project or a 200 line item unit price civil project would require several pages.

Use of technology

As for many other processes, estimating has been significantly affected by newer technologies. First, the use of electronic databases for historical costs is now prevalent in the industry. Both contractors and commercial data providers, such as RS Means, rely on databases to store cost and productivity information. Information is extracted from these databases to support estimating either through spreadsheets or other commercial or in-house estimating software. More recently, building and civil information modeling (BIM/CIM) tools allow automatization of the QTO step by extracting quantities directly out of three-dimensional models. However, technology enhancements should not ignore the nature of estimating as being both an art and a science. Therefore, contractors using technology tools continue to rely on expert estimators to review any result coming out of the "black box."

Project engineering applications

Some of the estimating tasks that may be assigned to project engineers, especially senior project engineers, include:

- Performing direct work QTOs,
- Pricing direct work by utilizing in-house databases,

Gateway Construction Company
2201 First Avenue
Spokane, WA 99205
509-642-2322

Summary Estimate (abbreviated)
Montana Highway 90 Overpass, Project #1732

Bid Item	Description	Quantity	Units	Material	Labor	Equip	Subs	Subtotal Cost	Pro-rated General Conditions	Pro-rated Fee	Total Bid	SOV Bid Unit Prices
1	Survey	1	LS		110000			110000	10560	12177	$132,737	$132,736.56
2	Restoration	1	LS	15000	15000			30000	2880	3321	$36,201	$36,200.88
11	Traffic Control	500	HRS				10000	10000	960	1107	$12,067	$24.13
17	Demo Striping	1200	LF				1200	1200	115	133	$1,448	$1.21
52	Concrete Pavement	70	SY	3000	12000	2500		17500	1680	1937	$21,117	$301.67
59	Bridge Deck Beams	9600	LF	230,400	85158	118139	283428	717125	68844	79383	$865,352	$90.14
67	Stair Walls	25	CY	2500	4000	3000		9500	912	1052	$11,464	$458.54
110	Sidewalks	600	SY	1000	4500	2300		7800	749	863	$9,412	$15.69
140	Conduits	230	LF				2300	2300	221	255	$2,775	$12.07
150	Catch Basins	5	ea				50000	50000	4800	5535	$60,335	$12,066.96
172	Pile Caps (1)	140	CY	30330	17289	7882		55000	5280	6088	$66,368	$474.06
200	Demobilization	1	LS	25000				25000	2400	2767	$30,167	$30,167.40
	Totals:			$2,250,000	$3,301,005	$950,000	$998,995	$7,500,000	$912,060 (2)	$1,097,940 (3)	$9,510,000	

Notes:
1 Although the assembly unit price from the pile cap recap sheet was $396/CY, the loaded unit price on the bid form will be $474 with markups
2 The general conditions are calculated as a standalone separate estimate, but then pro-rated equally @ 9.6% across the bid items on the SOV
3 Due to increased direct-labor risk, Gateway included a 13% fee spread proportionately across the bid form schedule of values (SOV)

Figure 17.6 Estimate Summary

- Calling subcontractors and material suppliers to request bids,
- Receiving subcontractor quotes on bid day, and/or
- Analyzing straightforward bid packages and recommending subcontractor choices to the lead estimator.

Eventually, if the firm is successful with landing the project, a senior project engineer will utilize the estimate to assist with subcontractor and supplier buyout, draft subcontracts and purchase orders, set up the cost control systems, and play a major role in cost coding invoices. All of these tasks have been presented in various chapters of this book.

Summary

Depending on the level of design development, various types of estimates exist to provide either an updated forecast of the project cost or a price for a construction contract. Estimating is a step-by-step, building-block process. In semi-detailed and detailed estimating, the first overview step determines whether the project will be pursued. The quantity take-off step is a compilation of counting items and measuring volumes. Pricing is divided between direct labor pricing, material pricing, and subcontract pricing. Labor cost is computed by using productivity rates and company-specific labor wage rates. Material and subcontract prices are developed most accurately by using competitive supplier and subcontractor quotations. The jobsite general conditions cost is a job cost and is schedule-dependent.

Home-office indirect costs are often combined with the desired profit to produce the fee. The fee calculation varies dependent upon several conditions, including company volume, market conditions, labor risk, and resource allocations. The project engineer can assist the project manager and/or estimator in a variety of estimating steps, but the final markups and final total bid are customarily determined by the officer-in-charge.

Review questions

1. What is the difference between a conceptual and a detailed estimate?
2. How does an estimator develop the work breakdown structure?
3. What is the greatest risk the contractor faces when developing a cost estimate?
4. What are two sources for labor productivity rates?
5. When would it be acceptable to utilize unit prices from published references for materials and subcontract items of work in the final bid, and when would it be ill-advised?
6. If an estimator is certain of receiving at least one or two subcontract bids on bid day, why should he or she bother with estimating that area of work?
7. When is a construction schedule developed in the estimating process and why?
8. What is an acceptable jobsite general conditions percentage? Why is this percentage generally higher for civil projects than commercial ones?
9. What is the difference between jobsite and home-office general conditions?
10. List two differences and two similarities between the development of a heavy-civil unit price bid and a typical commercial office building bid.
11. By utilizing the different wage rates presented in this chapter, which of the labor crafts has the highest wage percentage benefit add-on, and why do you think that craft?

Exercises

1. By using Figure 17.1 and our case study organization chart, assume a bid date three weeks from today and develop an estimating assignment list for each team member, including due dates. Include a "check person" for each major deliverable.
2. Prepare a project item list similar to the one shown in Figure 17.2 for our MXD project. Utilize CSI division 07 from the detailed estimate and schedule on the website to prepare this WBS to (a) the second level and then (b) the third level. Items that require decision regarding self-performed or subcontractor work should be noted.
3. In addition to the data shown in the one-page MXD summary estimate on the website, what other cost detail would a construction-savvy client look for from a contractor on a completely open-book project? If you were the client, list your top ten items.

References

Holm, L., Schaufelberger, J., Griffin, D., and Cole, T. (2005). *Construction Cost Estimating: Process and Practices* (1st ed.). Upper Saddle River, NJ, USA: Pearson Education.

Project Management Institute (PMI). (2016). PMI Lexicon of Project Management Terms, Ver. 3.1. Retrieved September 26, 2017, from the Project Management Institute (PMI) website: www.pmi.org/

18 Planning and scheduling

Introduction

Planning and scheduling—similarly to estimating, as discussed in Chapter 17—is a topic that has been covered in scheduling books and reference guides. The purpose of this chapter is to briefly discuss schedule development, focus on using the schedule as a project management tool, and examine the role of a project engineer in that process. As introduced in Chapter 6, time management is just as important to project success as cost, safety, and quality management. The key to effective time management is to carefully plan the work, develop a realistic construction schedule, and then manage the performance of the work. Schedules are working documents that need updating as the conditions change on the project. For example, if contract change orders modify some aspect of the scope, an adjustment to the schedule may be required. The schedule is a tool that aids the management of the project. Updates and control of the schedule were introduced in prior chapters. Schedule support in change orders and claims is discussed in Chapter 20.

Planning

Planning and scheduling are terms that many use synonymously, but we are going to separate the two into distinct phases and products. The schedule is the final product produced by an initial and thorough planning phase. Project planning is the process of selecting the construction means and methods and the sequence of work for a project. Planning must be completed before a schedule is developed. It starts with the assembly of all the information necessary to produce a schedule. Some of the steps in planning the project include:

- Developing a work breakdown structure (WBS): A WBS is a detailed hierarchy of project elements that defines the scope of work to be executed. An abbreviated example WBS for the I-90 overpass case study was shown in Chapter 17;
- Seeking input from key members of the project team, including foremen, superintendents, subcontractors, and suppliers;
- Making decisions with the superintendent regarding:
  - Site layout,
  - Direction of work flow and sequencing,
  - Means and methods, including forklift and cranes for hoisting and concrete forming, and
  - Safety;
- Making decisions regarding the mix of self-performed work versus that of subcontractors;
- Evaluating restraining factors, such as:
  - Crew makeup and sizes,
  - Material delivery dates,

- Weather, and
- Permits; and
- Relying on a complete estimate of self-performed work: The project plan is not prepared until the *estimate* has been completed, at least for the work that the contractor will perform with their direct workforce. Craft hours from the estimate are needed to determine activity durations for the schedule network. However, a preliminary or summary schedule is needed to estimate the cost of jobsite general conditions, as was discussed in Chapter 17.

In order for the plan to be "workable," it requires careful consideration of all of the activities above and buy-in from all of the construction team members, including:

- General contractor's officer-in-charge (OIC),
- Project manager (PM),
- Superintendent,
- Field foreman,
- Project engineer,
- Home-office staff support specialists, such as the scheduler, estimator, and quality and safety officers, and
- Subcontractors.

Once the project plan has been completed, it is time to develop a construction schedule. Individual tasks or activities to be accomplished were identified during the development of the WBS. The next step is to determine the sequence in which the activities are to be completed. This involves answering the following questions for each activity:

- What deliverables, permits, submittals, and contracts need to be in place before starting?
- What activities must be completed before this activity can start?
- What activities can be started once this activity has been completed?
- What activities can be performed concurrently with this activity?

Based on the answers to these questions, the schedule structure can be developed by using the logic among the activities. Even with all the computer sophistication available today, many contractors still develop their logic by hand on butcher paper or by rearranging colored sticky notes on a whiteboard. There is no exact format that the project plan takes; it can be an outline, notes, sketches, and scribbles on butcher paper, meeting minutes, or a roughed-out schedule without a timeline. This information is then provided to the scheduler to input into the selected scheduling software program. In some instances, the scheduler may be the same individual as the planner. On some projects, the project manager and/or senior project engineer, in conjunction with the super-intendent, may develop the schedule. The duration for each activity is determined by using the crew productivity factors from the cost estimate. At this point, the start and finish dates for each activity can be established, and the overall scheduled project completion date can be computed and compared with the contractual completion date. Also, all activities that cannot be completed late without delaying the project completion can be identified. These activities will be laid down along one or multiple sequential paths from start to finish; this is called the *critical path*. If the schedule shows completion later than the contract requires, some of the activities on the critical path must be accelerated or stacked to produce a schedule that meets the contract requirements.

Very few schedules are hand-produced anymore, although early drafts developed in the planning process may still be done this way. Today, most schedulers use computers to produce the final schedule product. The identification of the activities, the duration for each, and the sequence in which they are to be completed must be input into computer scheduling software for it to develop the schedule. The scheduling software does not plan the project, but it will plot the schedule and calculate the start and finish time for each activity. Computer-generated schedules allow the scheduler, project manager, and superintendent to quickly determine the effects of changes in schedule logic, delays in delivery of critical materials, or adjustments in resource requirements. Scheduling software is a valuable scheduling tool, but it cannot take the place of adequate planning.

Schedule types

Schedules take a lot of different formats, and similarly to planning, there is no "exact" form that the contract schedule should follow. Many project managers and schedulers have their personal preferences. Most schedules fall into one of two standard formats: bar charts and network diagrams. Bar charts relate activities to a calendar, but they generally show little to no relationship among the activities. Network diagrams show the relationship among the activities, and they may or may not be time-scaled on a calendar. Two diagramming techniques are used to represent network schedules. The first is known as the arrow diagramming method, in which arrows depict the individual activities and nodes depict milestones. The other is known as the precedence diagramming method, in which the activities are represented by nodes and arrows depict relationships between activities. Network diagram schedules are also referred to as "critical path" schedules, because the critical path is the longest path through the schedule and determines the overall project duration. Any delay in any activity on the critical path results in a delay in the completion of the project. All are good systems and may be appropriate for different applications. Schedules can also be prepared as different presentations, depending upon the anticipated use.

- *Contract schedules*, also known as formal schedules: These schedules will be provided to the client at the beginning and throughout the project delivery as required by the contract special conditions.
- *Summary schedules*: These schedules are often used for presentations, proposals, or management reporting. They are an abbreviated version of a detailed construction schedule, which may include 20–30 major activities. They are similar to the bar chart illustrated in Figure 18.1 for the Rose case study.
- *Detailed schedules*: These schedules are posted on the walls of meeting rooms or in the jobsite trailer. They are marked up with comments and progress. These schedules may also serve as submittals for the contract schedule requirements. A sample detailed network for the Rose case study is included on the companion website. Reliable Construction Company is using a "collapsed" schedule on that project for the detailed activities for the upper-floor mechanical, electrical, and plumbing (MEP) systems and finishes. The complete detailed schedule would be 80–100 items longer. As the contractor approaches the rough-in and finishes for the upper floors, the schedule will then be expanded for the to-go work and collapsed for the past and installed work.
- *Short-interval "look ahead" schedules*: These schedules focus on short-term field activities. They should be developed by each superintendent or foreman and each subcontractor each week. They may be drafted by hand in bar chart form, on an Excel spreadsheet, or with scheduling software such as Microsoft Project. The form or system does not matter. What is important is that they are produced by the people who are doing the work and that they are communicated and distributed to all involved. In this way, they are construction management tools. These schedules can be in two-, three-, or four-week increments, depending

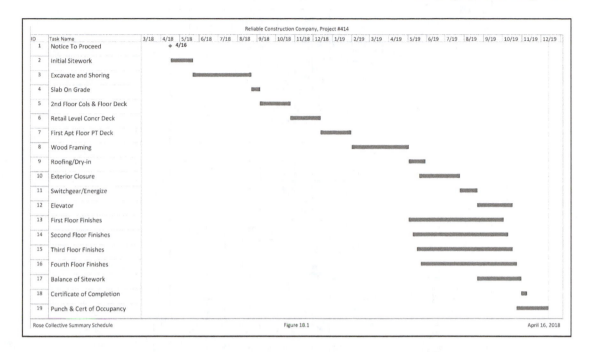

Figure 18.1 Summary Schedule

upon the job and level of activity. Some contractors simply print the activities for the next three weeks from the electronic master schedule without any additional detail or input from the superintendent. Figure 18.2 shows an example of a three-week schedule from our Rose case study project. As indicated, the schedule format is not as important as its author and content. This superintendent prepared the schedule by using Excel.

- *Mini-schedules*, *area schedules*, and *system schedules*: These schedules are intended to allow additional detail for certain portions of the work that could not be adequately represented in the project detailed schedule and have longer durations than the short-interval schedule.
- *Pull schedules*: These schedules are the product of *pull-planning* exercises that are one of the major lean construction tools being adapted from production industries. These schedules will be discussed later in this chapter.
- *Other schedules*: There may be many other specialized schedules on the project that include submittal, buyout, delivery, start-up, and close-out schedules. Many of these are discussed in other chapters in this book.

Traditional top-down scheduling

Top-down scheduling is frequently encountered on projects, especially public projects based on lump sum or unit price contracting that rely on low bidding for prime contractors and subcontractors. This approach to scheduling is not a good match with best value procurement, which was introduced in Chapter 5 for prime contracts. Best value subcontracting will be discussed in Chapter 19.

Top-down scheduling means that the client has dictated an end date and the general contractor (GC) has to meet it, at all costs, or will likely suffer liquidated damages. The GC will use the same process with their subcontractors. It is likely that the GC will have the schedule

RELIABLE CONSTRUCTION COMPANY

SHORT INTERVAL SCHEDULE

PROJECT: Rose Collective Mixed Use Apartment Building, Portland, OR

SUPERINTENDENT: Jason Campbell

DATE: Monday, 5/6/19

SHEET: 1 of 1

NO.	ACTIVITY DESCRIPTION:	S	S	6	7	8	9	10	S	S	13	14	15	16	17	S	S	20	21	22	23	24	S	S	COMMENTS
1	Roofing (Started 5/2)			X	X	X	X	X			X	X	X	X	X			X	X	X	X	X			Dry-In ECD 5/29
2	Continue Plumb R/I Floor 1			X	X	X	X	X			X	X	X	X	X			X	X	X					
3	Start HVAC R/I Floor 1						X	X			X	X	X	X	X			X	X	X	X	X			ECD 5/29
4	Start Elect R/I Floor 1													X	X			X	X	X	X	X			ECD 5/29
5	Start Fire Prot R/I Floor 1																				X	X			ECD 6/12
6	HM Door Frames in Hall, F1			X	X	X	X	X																	
7	Start MEP/Plumb Floor 2										X	X	X	X	X			X	X	X	X	X			Continues w/finishes
8	Start MEP/Plumb Floor 3													X	X			X	X	X	X	X			Continues w/finishes
9	Start MEP/Plumb Floor 4																				X	X			Continues w/finishes
10																									
11																									
11																									
12																									

Figure 18.2 Three-Week Schedule

produced by an in-house staff scheduler or maybe even the OIC, without necessarily involving the field management team. Often these schedules have little detail and are difficult to use on the jobsite as a construction management tool. Once the GC's superintendent receives a top-down schedule with only the contractual end date noted, he or she may, in turn, pass this same process down to their subcontractors and foremen and put the burden on them to achieve schedule success.

The schedule should be referenced in the contract as an exhibit, whether it is developed as a top-down schedule or collaboratively prepared as discussed below. It should be referred to in each sub-contract and purchase order by title and date. The project manager or senior project engineer may insert language into all subcontracts to place the subcontractors on notice that they are required to achieve the project schedule and that, if the general contractor determines that they are falling behind, the subcontractors will work overtime and/or increase crew sizes at their expense to catch up.

The contractual start date, or day one of the project schedule, is dependent upon receipt of four conditions from the client:

- Construction financing verification,
- Building permits,
- Signed contract, and
- Notice to proceed (NTP).

Signing of a contract with these four conditions means that the risk these conditions are not verified is on the project owner. On the other hand, signing of a contract that specifies a date, such as April 16, 2018, as the contractual start date results in the contractor accepting the risk. Therefore, many GC PMs will want to insert these four conditions into the contract, rather than just a date. On the other hand, most project owners and designers will want to use a specific date in both the commencement and completion articles of the construction contract. Another point of contention relates to the units used for project duration. Contractors usually prefer that completion is defined in work days after the issuance of the NTP and the other aforementioned requirements. On the other hand, project owners prefer using calendar days in conjunction with firm start and completion dates, which can burden the GC if some of the above requirements have not been accomplished.

Because the critical path does not include all the activities on a detailed schedule, the term "float" refers to additional time available for activities, or a string of activities, that are not on the critical path, before these activities move onto a new critical path. For activities outside the critical path, their duration can slip, either in start or end dates, but the project can still be completed per the contract schedule. However, it is also true that delaying these activities beyond their allowed float would make them become critical.

Who owns the float?

This is a critical schedule management issue that is usually specified in the contract. Usually, it is owned by the first party to use it. The client may claim that the float is theirs and use it to introduce new scope. The architect or engineer may use some of the float when responding to requests for information (RFIs) and submittals. Subcontractors and suppliers may use some of the float and either start their work late or deliver materials late. There have been numerous debates over float ownership, including many resolved by the courts; therefore, like top-down scheduling, the float is a contentious issue as well.

Collaborative approach

If they are interested in developing long-term team relationships with their clients and their subcontractors, contractors will use a more collaborative approach to schedule development. Also, collaborative schedule development is expected to result in a schedule that can be easily met by each party. Some of the methods to develop collaborative schedules include pull planning and buy-in from all the participants, from clients and designers to subcontractors and foremen.

Fast-track scheduling was a popular term utilized in the 1980s. Today, most construction projects are scheduled and built on a fast-track basis, which means that many activities are overlapping and concurrent, rather than being sequential. Very few schedules are completely linear, in that all preceding activities are 100% complete before the subsequent activity starts. Fast-track design and construction allows expedited start dates and turns the project over to the client faster, which reduces construction loan interest and improves developer performance. However, fast-tracking a schedule reduces leeway, so it is important that all team members are aligned around the schedule.

The schedule must be developed in sufficient detail to allow the project progression to be adequately measured, monitored, and corrected if necessary. We talked about the need for the estimate to be in sufficient detail to perform cost control in the preceding chapters. Too much detail, and a schedule will take on a life of its own and become too burdensome to monitor. The *80–20 rule* should be applied to schedules in the same way that it is applied to estimates and cost control work packages; 80% of the work or time is included in 20% of the activities. The project team should focus on the critical 20% tasks. Providing detail on a schedule is a way to reach out to foremen and subcontractors, and it improves their work planning efforts and the development of three-week look-ahead schedules.

Pull planning

> *Lean construction* extends the objectives of a lean production system—maximize value and minimize waste—to specific techniques and applies them in a new project delivery process.
>
> (Seed 2010)

The implementation of lean construction often equals a continual process for analyzing the delivery of a construction project to minimize costs and waste, while maximizing value. During pre-construction planning, the project team should examine planned processes for document flow, work flow, and material flow to minimize time delays and project costs.

Work flow is managed by the use of pull-planning schedules that start with project milestones and work backward to define what work must be completed to achieve each milestone. As needed, the project team will review the work accomplished and adjust the plan. Each foreman is asked to identify the work tasks that his or her crew will complete in the following week. Each individual foreman's commitment is connected to the commitments of the other foremen, which creates a network of commitments among the project team. The result is better short-term planning and control by having work flow between crews without interruption. This results in a collaborative approach to short-term scheduling. Pull planning does not replace traditional construction schedules but is used to improve the delivery of short-term assignments. The milestones selected to manage the pull-planning scheduling process come from the contract schedule. Pull-planning schedules are often developed on whiteboards by using multi-colored sticky notes. The schedules are also used to manage the delivery of construction materials to ensure that all needed materials are available for the crews but that excess materials are not ordered too early and then require extended storage on the construction site. Different colored notes are used for each subcontractor.

The client, designer, and city may also collaborate with their own colored notes on the GC's whiteboard.

Project owner and designer roles

Even if the project owner has an end date that the contractor must meet, they can still be collaboratively involved in the scheduling process. The owner will have the best contact with neighbors and the city, and good relations are essential for the GC, especially with respect to craft parking, deliveries, and work hours. The owner may also need to have partial occupancies and/or provide their own equipment for contractors to install and connect. Coordination and cooperation on these issues is essential for both parties' success.

The lead designer, whether an architect on a commercial project or a civil engineer on a heavy-civil project, will need to expedite permit release from the city. Partial permits or early design releases may give the contractor a jump-start on some of the early work packages. During construction, the design team can assist the construction team with schedule adherence by providing expedient and complete responses to RFIs and submittals. On negotiated building projects, the project owner and the architect may also invite the GC to assist with pre-construction services; this would entail early budgeting and constructability input to enhance quality, safety, and schedule performance.

Subcontractor and supplier involvement

Does the general contractor schedule the subcontractors or do the subcontractors schedule themselves? The same principle of commitment and collaboration applies to subcontractor-developed schedules as it does to superintendent-developed estimates and schedules. If the subcontractor develops, or at a minimum, inputs to the GC's schedule, and if it fits within the overall plan, they have made a commitment to achieve the schedule. However, if the general contractor dictates in a top-down fashion to the subcontractors when and how long they will be on the job, the subcontractors may accomplish the task, but if they do not finish on time, they will always have the excuse that they did not get the opportunity to provide input.

Field foremen buy-in

Project managers, project engineers, and superintendents do not actually "build" the facility or infrastructure. Rather field construction is accomplished by the many construction craftsmen and foremen. The importance of field involvement in safety, cost, and quality control was introduced in Chapters 7, 8, and 9, respectively. The foreman's work package is a combination of estimate and schedule input to allow them to develop a workable plan for an area or system of work, such as steel column erection. Weekly foremen meetings should be held on Monday or Tuesday of each week, and each of their three-week look-ahead schedules should be copied and distributed to each other, including direct craft foremen and subcontractors. A detailed construction schedule will assist with the development of these short-interval schedules. Logistical issues, such as the scheduling of deliveries and use of hoisting, are worked out at these meetings with a collaborative approach. It takes collaborative involvement from many team members to successfully achieve a schedule and build a quality facility or infrastructure.

The jobsite project team should be actively involved in developing the construction schedule, because they will have the responsibility of completing the project in the desired time. This provides buy-in from the project leadership. If they agree with the schedule, they will do everything they can to make it happen.

Use of technology

As for estimating, scheduling heavily relies on the support of technology. The use of scheduling software, such as Microsoft Project or Oracle Primavera P6, is now prevalent in the industry. These platforms are often integrated with handheld devices that can provide schedule updates from the field. More recently, specific building and civil information modeling (BIM/CIM) tools allow contractors to add a time dimension to three-dimensional models; these tools are often defined as four-dimensional (4D) BIM/CIM tools. These 4D tools are used to assign times of installation to different geometric components and to run temporal simulations of the installation process. The main usefulness of these tools is to visualize the planned construction process so that potential issues can be identified before they arise in the field and changes can be proactively incorporated.

Project engineering applications

New project engineers will not "plan" or "schedule" a project per se. Their role, as with many other tasks described in this book, is one of support. However, senior project engineers will have opportunities to produce mini-schedules and input the workable plan into the computer to produce drafts of the detailed contract schedule. Some of the ways project engineers may become involved in planning and scheduling include:

- Drafting a jobsite layout plan with input from the superintendent,
- Calling subcontractors for durations, manpower, and delivery dates,
- Assisting the superintendent with the development of three-week short-interval schedules,
- Developing specialty, area, or subcontractor schedules,
- Implementing schedule revisions in the computer software (see Chapter 16), and/or
- Attending as many meetings chaired by the superintendent and project manager as possible, including pre-construction meetings and foremen coordination meetings.

Summary

Just as a hammer and saw are important tools of the carpenter, schedules are construction management tools for the superintendent, project engineer, and project manager. Schedules are contract documents and are important tools for the client and design teams as well. Proper planning of the project and the schedule, with input from the relevant personnel, such as the superintendent and the subcontractors, is key to developing a useful construction management tool. Schedule development begins with proper planning that considers many variables, such as deliveries, logic, manpower, and equipment availabilities. Schedules can take on many different formats, but what is important is that they are developed by those who will utilize them to plan and communicate their work activities. Top-down schedules involve outside mandated target dates without the input of all project team stakeholders. Collaborative schedules that elicit input and buy-in, all the way from the client through the subcontractors and foremen, are found to be more successful. This is similar to "participative leadership," which will be introduced in Chapter 22.

Review questions

1. What is the difference between planning and scheduling?
2. Which members of the construction team input to the project plan?
3. What is the difference between a bar chart schedule and a network diagram schedule?

4. Who should develop a three-week look-ahead schedule, and why them?
5. Are schedules contract documents?
6. When does day one of the contract schedule occur?
7. How many activities should be on the project schedule?
8. How does the 80–20 rule apply to developing schedules?
9. List two distinct differences between top-down and collaborative scheduling.

Exercises

1. Search for examples of schedules using both arrow diagramming method and precedence diagramming method techniques. Then, draw a three-task schedule that includes: (a) site work, (b) concrete slab, and (c) shed construction by utilizing both techniques.
2. How many work items or elements should be included in the work breakdown structure?
3. Develop a short-interval schedule for the construction of the slab on grade (SOG) and foundation work for the Rose case study. Identify the activities and logic, and take a shot at determining reasonable durations for the activities.
4. By utilizing the detailed schedule for the Rose case study on the website, develop a schedule unique to the electrical contractor. Include the work of other trades if it affects the electrician's work.
5. Prepare a five-point argument to your CEO for why he or she should not develop the schedule themselves for your project, if you are the GC's PM on the Rose case study project.
6. Have you ever been involved in a project that had a top-down scheduling approach? Was it successful? Why or why not?
7. Have you ever been involved in a project that had a collaborative scheduling approach? Was it successful? Why or why not?
8. A few milestones have been shown on the Rose detailed schedule on the companion website. Prepare a separate milestone schedule for management reporting. Include the milestones shown and add five or so additional milestones that you feel would be important to the Reliable Construction Company, the developer, the architect, and/or the City of Portland.

Reference

Seed, W. R. (2010). *Transforming Design & Construction: A Framework for Change* (1st ed.). Arlington, VA, USA: Lean Construction Institute (LCI).

19 Procurement management

Introduction

Very few general contractors (GCs) self-perform all the work on a construction project. In the building sector, most GCs use subcontractors to execute on-site construction activities and suppliers to provide the materials. A typical general contractor would subcontract up to 80–90% of the work on building projects. Heavy-civil GCs usually self-perform a larger portion of the work, but they also rely on subcontracting to deliver some specialty components or to share risk with another heavy-civil GC. Subcontractors are, therefore, important members of the GC's project delivery team and have a significant impact on the project success or failure. The relationship of the subcontractors to the other members of the project delivery team was illustrated in Chapter 3. Because subcontractors have such a great impact on the overall project success, they must be selected carefully and managed efficiently. There must be mutual trust and respect between the superintendent and the subcontractors' field management teams, because each can achieve success only by working cooperatively with the other. Consequently, project managers, project engineers, and superintendents find it advantageous to develop and nurture positive, enduring relationships with reliable subcontractors. The GC's project management team must treat subcontractors fairly so that they remain financially stable.

This chapter describes subcontractor and supplier selection and management. Selection of quality subcontractors is essential if the GC's jobsite project team is to produce a quality project on time and within budget. Poor subcontractor performance will also reflect negatively on the GC's professional reputation and their ability to secure future projects. Once the subcontractors have been selected, contract documents are executed to secure scopes of work, as well as terms and conditions of the agreements. Subcontractor and supplier management is an integral part of construction management. The GC's project superintendent manages the field performance of the subcontractors, with help from foremen and field engineers, and the project manager and project engineers manage subcontract and purchase order documentation and communication. We are covering both subcontractors and suppliers in this chapter and may, for convenience, only refer to one or the other, but the concepts included here apply to both groups in most cases.

Risk management

Two of the reasons GCs use subcontractors are to reduce exposure to some risks and to have access to specialized skilled craftsmen and their equipment. Whereas risk management will be discussed at large in Chapter 21, this section specifically describes how GCs use subcontracting as a risk management approach.

One of the major risks in contracting is to accurately forecast the direct cost of labor required on site. By subcontracting significant segments of work, the general contractor can transfer much of the *labor risk* to subcontractors. When the GC's estimator asks a subcontractor for a price to perform a specific scope of work, the subcontractor bears the risk of properly estimating the labor, materials, and equipment costs. Craftspeople experienced in the many unique and specialized trades required for major construction projects are expensive to hire and are generally used on a project site only for limited periods of time. It would be cost-prohibitive for GCs to employ individuals in all of the different specialized trades as part of their own full-time labor force.

Subcontracting reduces the GC's exposure to some risks, but it does not eliminate all risks. The general contractor's project team gives up some control when working with subcontractors. The *scope* and terms of each subcontract agreement define the responsibilities of only that subcontractor. If some aspect of the work is inadvertently omitted, the project manager, project engineers, and superintendent are still responsible for ensuring the overall contract requirements are achieved. Specialty contractors are required to perform only those tasks that are specifically stated in the subcontract agreement. For the GC, the sum of the parts—or in this case, subcontractors—must equal the whole. Consistent *quality control* may be more difficult with subcontractors, particularly with regard to the quality of field workmanship. Clients expect to receive a quality project and hold the GC accountable for the quality of all work, whether it was performed by the general contractor's own crews or by subcontractors.

Subcontractor *bankruptcy* is another risky aspect of subcontracting, which can be minimized by good pre-qualification procedures and timely payment for subcontract work. *Scheduling* of subcontractor work is often more difficult than scheduling the general contractor's crews, because the subcontractor's craftsmen may be committed to other projects. General contractors will argue that *safety* practices among subcontractors are not as effective as those used by their own craftsmen, which presents additional challenges for the GC's superintendent. Additional discussion of construction risk-management is covered in Chapter 21.

Types of subcontracting

Several approaches to subcontracting are used in the industry, depending on the services to be subcontracted. This section describes them and their contractual frameworks, as shown in Figure 19.1.

Traditional subcontracting reflects the traditional contracting approach, where the project owner contracts out design and construction services to two different entities, the lead design firm and the prime contractor. Under this approach, the lead design firm then contracts out part of their scope to design sub-consultants. Similarly, the prime contractor contracts out part of their scope to subcontractors. Under this approach, there are no mandated functional relationships between the design sub-consultants and subcontractors. This approach also relies on the assumption that the sub-consultants would be able to provide a buildable design. Often, however, this assumption is unmet and communications between subcontractors and sub-consultants need to occur to refine the design to the point of buildability. Contractually, these communications need to go through the prime contractor–owner–lead designer chain (Osmanbhoy 2015; pp. 54).

However, design knowledge for some specialties has evolved to the point that specialty contractors and sub-consultants share the expertise of producing a buildable and functional design. To accommodate these situations, the traditional approach has been adapted to establish a functional *design–assist* relationship between sub-consultants and subcontractors, which requires the subcontractor to assist the sub-consultants during the design phase. Under this approach, the sub-consultant retains the design liability, while seeking assistance during design from the subcontractor (Osmanbhoy 2015; pp. 56–57).

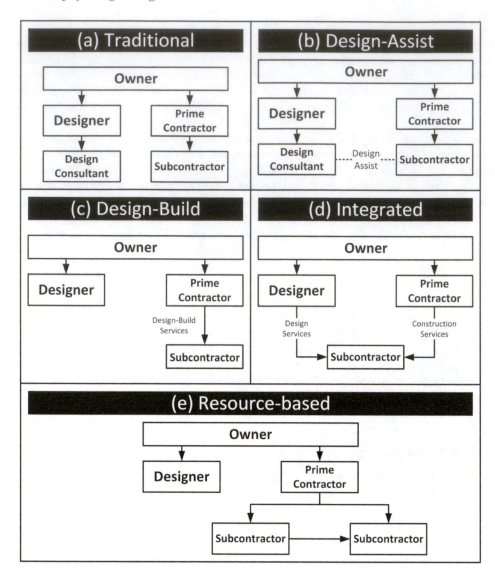

Figure 19.1 Types of Subcontracting

Because some subcontractors have evolved to the point of being able to design and build their scope, *design–build subcontracting* replicates the design–build delivery method at the subcontracting level. Under this approach, the design–build subcontractor will design and build its scope within the boundaries provided by the lead designer, while retaining an individual contractual relationship with the prime contractor (Osmanbhoy 2015; pp. 58–59).

To achieve a similar approach, *integrated subcontracting* allows a highly skilled subcontractor to enter into two separate subcontract relationships, one with the lead designer for design services and one with the prime contractor for construction services. Though experimental in the industry, this approach is intended to achieve similar results to design–build subcontracting, while allowing for direct communication with the lead designer (Osmanbhoy 2015; pp. 60–65).

Finally, *resource-based subcontracting* is sometimes used by a subcontractor to contract out some of its scope to another project subcontractor. This approach is mostly used to create efficiency in terms of resource use. For instance, a subcontractor that would need to mobilize special equipment to perform a task could use this approach to subcontract the task to another subcontractor that has already mobilized the same equipment to the project (Osmanbhoy 2015; pp. 66–67).

Subcontractor selection

General contractors who focus on proactive quality management pre-qualify subcontractors and suppliers before asking them to submit a bid on a project. Historically, too much emphasis has been placed only on price and not enough on quality and experience, because of the prevalent use of low bidding. When the owner's procurement allows for it, the project team should select the "best value" subcontractors, which requires evaluation of subcontractors' costs, schedules, quality, and safety performances, rather than just selection of the lowest bid. A long list of subcontractors and suppliers for each category of work should be developed by the GC. For each of the listed subcontractors, the following issues should be evaluated:

- Subcontractor experience,
- Subcontractor project managers and superintendents,
- Financial strength,
- Specialized equipment,
- Safety record, and
- Past project performance with respect to schedule and quality.

A request for qualifications (RFQ) can be utilized to gather this information from prospective subcontractors, similarly to the RFQ used by owners to shortlist GCs. Additional information should be solicited from other experienced in-house managers, and personal interviews may be required before selecting the short-list set of pre-qualified subcontractors invited to submit proposals. On projects based on negotiated contracts, owners and designers should be invited to participate in reviewing the qualifications of prospective subcontractors. The GC's estimating team will only use best value subcontractors as potential team members.

Once the GC has selected the top three to five subcontractors for each category of work, they are invited to submit a proposal or bid. Requests for proposals (RFPs) are typically used for negotiated projects, and invitations to bid (ITBs) and/or requests for quotations (also referred to as RFQs) are used for lump sum bid or heavy-civil unit price scopes of work. Most estimators solicit bids by using written RFPs, ITBs, and/or RFQs to ensure that each prospective subcontractor understands the specific project requirements.

Subcontractors and suppliers have historically phoned in their bids to the GC, but today many bids are also submitted electronically. Bids received from one subcontractor should not be disclosed to another to drive down prices. This practice, known as *bid shopping*, is unethical and, on some public projects, also illegal. Subcontractors expect GCs to treat their bids as confidential information until the subcontracts are awarded. GC estimators who bid shop for subcontractor bids will find fewer firms willing to supply competitive bids on future projects. Thus, the American Society of Professional Estimators (ASPE) identifies bid shopping, defined as "... when a contractor contacts several subcontractors of the same discipline in an effort to reduce the previously quoted prices" (ASPE 2011), as being in direct violation of their Code of Ethics. Another unethical practice is *bid peddling*, which ASPE defines as "... occur[ring] when a subcontractor approaches a general contractor with the intent of voluntarily lowering the original price below the price level established on bid day. This action implies that the subcontractor's original price was either padded or incorrect. This practice undermines the credibility of the professional estimator and is not acceptable" (ASPE 2011).

After bids are received from the subcontractor short list, they are evaluated to select the best contractor for each category of work. A *subcontractor analysis form*, or *bid tab*, can be used to assist in subcontractor selection. To ensure each subcontractor's cost proposal is realistic, the project manager or estimator should have previously developed an order of magnitude cost estimate for each category of work. If subcontractor or supplier bids are exceptionally low or high, this often indicates that the bidder might:

- Be uncertain regarding scope,
- Have felt the documents were unfavorable to their style of work, or
- Have made an estimating mistake.

A high bid often also indicates that the subcontractor is too busy with other work and not interested in this project. All discrepancies regarding supplier or subcontractor scopes of work should be resolved before a purchase order or subcontract is executed.

Selecting best value specialty contractors for each sub-trade category is essential for the GC's success. Quality specialty contractors have good safety records, experienced craftspeople, proper equipment, and adequate financial capability to complete the project to the expected standards without experiencing financial problems. Some project teams select subcontractors simply based on price, which is often a requirement for public lump sum bid and heavy-civil unit price projects. Selection based solely on low price may lead to problems on the project with respect to quality, safety, or schedule controls. When it is allowed by the procurement approach, the goal of a GC should be to select the subcontractors that provide the best value to the project. On some privately financed projects, the client reserves the right to approve the final subcontractor selection. Although qualification-based selection of subcontractors is highly recommended, some projects, especially public works projects, require an open bidding system to be used for the selection of subcontractors. Purchase orders and subcontracts awarded solely due to the lowest bid are not necessarily the best value.

Performance and payment bonds may be required of subcontractors. If the project owner does not require bonds, the general contractor may not require subcontractor bonds, especially if the subcontractors are also selected based on their technical competence, reputation, and financial strength. Independent from the owner's requirements, general contractors sometimes require bonds on subcontracts exceeding a certain value, such as $200,000. However, bonds will generally be required by the GC's surety if the contract with the client required it to be bonded. Bonds are not free, and for subcontractors, they can range from 1% to 2% of the cost of the work. If the owner required subcontracts to be bonded, the surety fees would be included in each subcontractor's bid. Bonds protect the GC against subcontractor default, bankruptcy, and liens, but they do not guarantee a smooth project. The subcontractor's bonding surety guarantees subcontractor performance to the general contractor, whereas the general contractor's performance is guaranteed by their surety to the client.

Subcontract management

The receipt and analysis of subcontractor quotes or proposals are sometimes performed by the estimator or a procurement specialist, but often the project manager, project engineer, and superintendent will weigh-in with their recommendations. Once the prime contract is executed, the general contractor will assemble a team to buy out all selected specialty contractors by finalizing and executing their subcontracts. A clear scope of work must be established for each of the selected specialty contractors, agreed upon, and delineated in subcontracts and purchase orders. The price is entered on the subcontract, and the agreement is signed both by the GC's chief executive officer (CEO) and the specialty contractor. The procedures used are similar to those discussed in Chapter 5 for executing the prime contract between the project owner and the general contractor. After the contracts are executed, then it is the jobsite team's responsibility to manage what may be up to 100 second-tier subcontracting companies.

Scope

The work breakdown structure (WBS) developed by the GC's estimator identifies the items that are intended to be self-performed by the general contractor and those that are to be subcontracted. After the WBS is ready, a subcontracting plan is prepared that identifies which scopes to include in each subcontract, essentially dividing up the whole subcontracted scope among each intended subcontract. The *subcontracting plan* will become the basis for the project manager (PM) and project engineers (PEs) to draft each unique subcontract and purchase order. The subcontract must clearly define what is included and what is excluded. A well-defined scope of work is essential to ensure specialty contractors understand what is expected of them. Poorly defined scopes of work lead to conflicts and may result in schedule delays, change orders, and disputes. The subcontract agreement must state clearly the exact scope of work to be performed, along with specific inclusions and exclusions, as well as reference all contract drawings and specifications.

Clear and concise language should be utilized for subcontract scopes, with emphasis on inclusions and exclusions. Terms such as "provide" and "install" should be used to clearly define the subcontractor's responsibilities. Suppliers only provide material and may include off-site fabrication labor, whereas subcontractors also provide on-site labor. Suppliers receive a *purchase order*, and subcontractors receive a *subcontract agreement*. Subcontractors may fall into one of the following three additional levels of specialization:

- Subcontractor provides labor and equipment only and installs materials provided by another supplier or through the general contractor;
- Subcontractor provides labor, equipment, and their own materials, which may have been purchased off the shelf, bought from a third-tier supplier, or fabricated in the subcontractor's yard or warehouse; or
- Subcontractor provides labor, equipment, and materials as indicated above but also performs the design related to their work and may or may not obtain their own permits. These subcontractors are referred to as design–build subcontractors, and this type of subcontracting was described earlier in this chapter. Examples include mechanical, electrical, landscaping, and curtain-wall subcontractors, among others.

The general contractor must ensure that all work items to be subcontracted are included in a subcontract scope of work, and nothing is omitted that could later result in unanticipated expenses. Conversely, the buyout team must also ensure that individual work items do not appear in more than one subcontract or purchase order.

Subcontract documents

Subcontract agreements clearly define the understanding between the GC and each subcontractor, as purchase orders do with suppliers. These agreements should be based on the same drawings and specifications referenced in the prime contract with the project client. Once the scope of work has been established, the specific terms and conditions are defined for each subcontract, to establish the operating procedures the jobsite project team intends to utilize with each purchase order and subcontract agreement. These contracts must define:

- Scope based on contract documents,
- Price,
- Schedule and/or delivery dates,
- Cleanup and safety,
- Daily diaries and three-week schedules,
- Hoisting,
- Insurance and bond requirements,
- Payment request and change order procedures,
- Inspections, warranty, and close-out expectations, and
- Process for claims, dispute resolution, and termination.

Standardized subcontract and purchase order documents are available for use by GCs from the American Institute of Architects (AIA) and ConsensusDocs. Project managers and project engineers should choose one of these common documents because they are used widely in the industry and understood by all parties. The advantage of using standardized documents is that they have been tested in court and found legally sound. GCs who decide to draft their own and do not use one of these standard documents should consult with legal counsel first.

Jobsite management

The GC's project team, which typically includes the project manager, project engineer, and the superintendent, must work together to schedule and coordinate subcontractors' work to ensure the project is completed on time, within budget, safely, and in conformance with contract quality requirements. Because most of the construction work is performed by subcontractors, efficient management of their work is critical to the GC's ability to control costs and complete the project on time. After bid and award, but before subcontractors are allowed to start work on the project, they are required to provide certificates of *liability insurance*, similar to those required of the GC by the project owner. Coverage requirements will be dictated in the subcontract agreement.

The GC's superintendent must orchestrate all independent subcontracting firms to work together as a cohesive project delivery team, which may include more than 100 firms. This requires an understanding of their concerns and proper work sequencing to ensure their success. It is essential that the superintendent establishes a cooperative relationship with the subcontractors and their foremen by conducting weekly coordination meetings. Frequently, open communications coupled with mutual goals will help foster subcontractor relationships that are built on trust. Leadership is an attribute of successful superintendents, as will be discussed in Chapter 22. The superintendent should require each subcontractor to submit a *subcontractor's daily report* of their activities. This provides a daily record of each subcontractor's progress and any obstacles encountered in performing the work. The GC's requirement of the subcontractors to submit daily reports must also be included in each subcontract agreement.

The GC's success on any project is also dependent on a workable and detailed *schedule*. The superintendent is responsible for coordinating start and completion dates with each subcontractor. They need adequate time to obtain equipment and materials and mobilize their crews. This means the superintendent must provide adequate notice to each subcontractor regarding the scheduled start date for their portion of the work. Timely notice is fundamental to building good relationships with subcontractors. The notice should also indicate the scheduled completion date to preclude interfering with the work of the following subcontractors. Pull-planning scheduling was discussed in Chapter 18 as a means of collaborative team efforts.

Subcontractors are required to understand their quality control commitments and other contractual obligations before being allowed to mobilize and proceed with work. The GC's project superintendent conducts *pre-construction meetings* with subcontractors before allowing them to start work on the project. The client and designer should be invited to participate in all subcontractor pre-construction meetings, especially on private negotiated projects. Often *mock-ups* are required for exterior and interior finishes. This allows the superintendent and the designer to evaluate the work and establish a standard of workmanship. Project site *cleaning policies* should also be discussed at the pre-construction conference. Subcontractors are required to clean up after themselves when they have completed their portion of the work.

Before each subcontractor starts their portion of the work, the superintendent must make sure that the jobsite is ready for them. If subcontractors arrive on a site that is not ready, it causes hardship to them and can result in lost time and potentially increased costs. Multiple subcontractors often work concurrently on the project site, and the superintendent must ensure that their workflows are compatible and that all parties *communicate*. The superintendent is also responsible for resolving any *conflicts* between subcontractors.

The GC's project team needs to manage the work of all subcontractors, so that their work conforms to contractual requirements. The superintendent must ensure subcontracted work conforms to *quality requirements* specified in the prime contract. Subcontractors must not be allowed to build over work that has been improperly installed by a previous subcontractor. The superintendent should walk the project when each subcontractor finishes their portion of the work to identify any needed rework. An *inspection report* similar to the one illustrated in Chapter 9 should be used to document the results of each inspection. This is better than waiting until the end of construction and trying to get several subcontractors to return and do rework. Early identification of deficiencies through a pre-punch list process is an important ingredient of an active quality management program, as discussed in Chapter 9.

An additional requirement for all subcontractors is a clear and strict definition of their *safety responsibilities*, regarding not only their own craftsmen but also those of other subcontractors working around them on the site. Project engineers often assist the superintendent with several facets of subcontractor coordination. Even though most of the work on a jobsite is being performed by subcontractors, the general contractor and its superintendent are still responsible for all personnel working on site.

Similarly to the GC and suppliers, subcontractors often require document clarifications from the design team regarding materials and scope. A *request for information* (RFI) procedure should be established to document subcontractor questions and responses. Some questions can be answered by the project superintendent or an experienced project engineer, but most will require resolution from the design team. RFIs and an RFI log should be used to expedite the processing of subcontractor inquiries; this process is customarily managed by the project engineer. Examples of an RFI and RFI log were introduced in Chapters 10 and 11.

Subcontractor and supplier scopes of work require modifications when conditions are different than those they contracted to. All such modifications should be documented as subcontract *change orders* to clearly identify the changes in scope and the impact on the subcontract price. Sometimes changes affect multiple subcontractors, and their contractual agreements must all be modified accordingly. Prior to negotiating the cost of a change order, the project manager or senior project engineer sends a change order request to the subcontractors that are potentially impacted. All subcontractor-generated change orders should be issued in writing on the subcontractor's letter-head. Subcontractor- and supplier-generated change order proposals (COPs) should be processed and logged, similarly to the process discussed in Chapter 20.

Subcontractors submit monthly *requests for payment* to support the GC's pay request process with the client. Once the project engineer and project manager approve a payment request and the client has paid the GC, payment should be made to subcontractors and suppliers. Payments should be made within the timeline established in the contract agreements. Most subcontracts contain a provision that the subcontractor will be paid for their work once the project owner pays the general contractor for the same portion of work. This is known as the "pay-when-paid" clause. Such contracts place significant financial burdens on subcontractors if project owners fail to make timely progress payments to general contractors. There are other standard subcontracts that contain a provision that the general contractor will pay the subcontractor for work completed within a "reasonable time," whether or not the owner has paid the general contractor for the work. These contracts place the total risk of client non-payment on the general contractor. As the subcontractor has performed its responsibilities under the terms of the subcontract, many in the construction industry believe payment should be made whether or not the client has paid the general contractor; different states have different laws with respect to this issue. Regardless, subcontractors should be paid on time to avoid cash flow problems and to ensure their financial solvency. Prior to the general contractor's PM and PE processing the final payment request, the subcontractor should also submit a final *lien release*. The GC's pay estimate process is introduced in Chapter 22.

Supplier management

Much of the previous discussion on subcontractor management also applies to suppliers, with only a few exceptions. For convenience, the similarities will not be repeated. Instead, this section highlights just a few of the differences. Suppliers provide materials or equipment only. This possibly includes specialty off-site fabrication labor, such as structural steel or cabinets, with the prefabrication taking place in the supplier's factory or yard. Suppliers do not have any *on-site labor*, unlike subcontractors. Therefore, suppliers do not provide liability insurance and rarely provide performance and payment bonds.

Suppliers receive a *purchase order* (PO), whereas subcontractors execute a subcontract agreement with the GC. There are two major types of purchase orders: a long form and a short form. The long-form PO is usually authored by the project manager, an experienced project engineer, or a staff procurement specialist in the main office. This PO is for materials with larger values, those with longer lead times, and/or those that require off-site fabrication and submittal approval. An example of a long-form PO for our heavy-civil I-90 highway overpass case study project is included as Figure 19.2. Suppliers who will typically receive a long-form PO include those providing:

- Concrete reinforcement steel (rebar), when installed by a separate subcontractor or the GC;
- Redi-mix concrete;
- Pre-cast concrete;
- Structural steel, when installed by a separate subcontractor or the GC;
- Wood framing, including fabricated lumber like glue-lam beams and trusses, when installed by a separate specialty subcontractor;
- Windows, if not a curtain-wall system, which will be subcontracted;
- Doors, frames, and hardware; and
- Specialties, such as toilet accessories, fire extinguishers, lockers, signage, and corner-guards; most of these are installed by the general contractor's carpenters.

MATERIAL PURCHASE ORDER CONTRACT

Gateway Construction Company
2201 First Avenue
Spokane, WA 99205
509-642-2322

TO:	Idaho Precast Box 455 Highway 101 Coeur d' Lane, ID 83814	PURCHASE ORDER NO.: <u>1732.12</u> DATE: <u>April 21, 2018</u>
FOR:	Pre-Cast Concrete Beams	PURCHASE ORDER VALUE: <u>$230,400.00</u>
SHIP TO:	I-90 Jobsite, Missoula, MT	SHIP VIA: <u>Your Truck</u>
SCHEDULED DELIVERY DATE:		July 1, 2018, Pending confirmation with project superintendent

SPECIFIC INCLUSIONS:

1. Supply concrete and reinforcement steel and formwork and other means and methods necessary to fabricate all concrete pre-cast beams according to Montana State specification section 035600, contract drawings, and bid item # 059.
2. Bid item 59 currently scheduled for 20 beams at 48 feet long each for a total of 9,600 LF at the Idaho Precast bid amount of $240/LF for a total purchase order amount of $230,400.00.
3. Fabricator is responsible to prepare and submit fabrication drawings for approval by the State engineer no later than May 5, 2018. Fabricator shall not begin fabrication of pre-cast beams until the State has approved the fabrication drawings. Any earlier fabrication will be at supplier's risk.
4. Fabricator is responsible to prepare installation drawings, for installation by others, no later than June 15, 2018.
5. All shipping and transportation costs, including cross-state taxes, are the responsibility of the fabricator.

SPECIFIC EXCLUSIONS: Truck unload and installation by others

This Purchase Order is subject to additional terms and provisions printed on the reverse hereof.

SELLER:	BUYER:
By: <u>Jim Bergerson</u> Title: Idaho PC CEO	By: <u>Rachel Radcliff</u> Title: GCC Project Manager
Purchase Order Prepared By:	Fritz Johnson, GCC Project Engineer

Figure 19.2 Long-Form Purchase Order

The short-form purchase order is, by definition, shorter than the long-form PO. An example of a short-form PO for our I-90 case study is included as Figure 19.3. The long-form PO will have much more detail, fine print, or boilerplate terms and conditions. The short-form PO is usually authored by the superintendent, often with assistance from the field engineer or project engineer. The materials purchased on the short form are usually "off-the-shelf" and ready for pick up at a

PURCHASE ORDER

Gateway Construction Company
2201 First Avenue
Spokane, WA 99205
509-642-2322

To: Grizzly Aggregates
Route 5, Box 643
Missoula, MT 59801

P.O. No.:   **1732.05**
Date: 4/15/2018
Job Name: I-90 Overpass

Attention: G. Adams
Phone No.:   406-655-1422

Job Address: Missoula, MT

When Required: As Needed			Ship Via: Your Truck, 5 CY Minimum		
No.	**Quantity**	**Unit**	**Description**	**Unit Price**	**Amount**
1	Min 5	CY	Drain Gravel	$30/CY	$
2	Min 5	CY	Sand Bedding	$20/CY	$
3	Min 40	LF	16" Diameter PVC Culvert	$12/LF	$
4	Min 5	CY	Landscape Topsoil	$40/CY	$
				TOTAL	$

Purchase Order Instructions: Material only to be ordered by persons noted above. All deliveries are to be scheduled. Driver is required to wear a hard-hat, orange vest, and construction boots

Requested By:	Approved By:	Purchased By:
Randy Buckwater	Randy Buckwater	Firtz Johnson
Superintendent	Superintendent	Project Engineer

Figure 19.3 Short-Form Purchase Order

local material supply warehouse. The value of materials purchased with the short-form PO is often limited to a certain value, say $200 or less, or for as-needed unit price items with no set total lump sum value. Larger value materials are purchased with the long form. Some examples of materials purchased with the short-form PO include nails, caulking, glue, small tools, short-term equipment rental, gravel and dirt, smaller orders of lumber, rebar tie-wire, and rebar hi-chairs.

Most jobsites have limited laydown area, and even if they were not limited, it may be unwise to have materials delivered too early to a site because of potential theft or damage from weather. The general contractor will schedule material deliveries for only when they are needed, also known as *just-in-time delivery*. In the case of subcontractors who install their own materials, they are more in tune with when materials are needed for installation, and they have a vested interest to participate in the management of that process. The general contractor's project superintendent plays a key role in allocating space on the jobsite for the temporary storage of materials waiting for installation. The materials need to be out of the way of other contractors' work but close enough to allow an efficient installation process and avoid costly double handling of materials.

Off-site prefabricated materials and efficient use of jobsite laydown are additional examples of lean construction techniques.

Suppliers are, of course, motivated to receive payment for their product, which can usually only occur after the materials have been received. Suppliers of materials that require extensive off-site fabrication, such as structural steel or kitchen cabinetry, may request payment for materials held in their yard or warehouse. There are complications with paying for *off-site stored materials*, such as property insurance and bonding, but sometimes it is an advantage for both the GC and the project owner to authorize these payments. In such a case, it is recommended that the GC's project manager or an experienced project engineer visit the vendor's place of business and inspect the materials, with backup photograph documentation; this may be a client or bank requirement as well.

As suppliers do not have any on-site labor, *retention* is not held from their monthly invoice, as is the case with subcontractors. Some suppliers may offer the GC a *discount* if payment is made within ten days of receipt of the invoice. This usually amounts to a percent or two of the value of the materials. Discounts apply to short-form POs more than long-form POs. This is because larger suppliers are likely subject to the same pay cycle that subcontractors are subject to; that is, they are paid monthly after receipt of payment from the client to the GC.

Suppliers do not play as heavy of a role during *close-out* as subcontractors. Because their materials should have been inspected and approved before unloading and certainly before installation, they are not involved in punch lists as much as subcontractors. The major suppliers still provide input to the operation and maintenance (O&M) manuals, including warranties, but they do not produce as-built drawings. Suppliers will not have taken out specialty permits, like some subcontractors do; therefore, these will not be included in the O&M manuals either. Because suppliers are not subject to retention, many of them have been paid in full and their final lien releases have been collected before the GC's close-out phase.

Project engineering applications

As discussed in Chapter 17, project engineers first help in the procurement process by assisting the estimator and project manager with subcontractor and supplier bid solicitation and by receiving those bids on bid day. Although the choice of the "best value" subcontractors and suppliers will ultimately be the responsibility of the project manager and the superintendent, the project engineer may have opportunities to attend the pre-award meetings and complete relatively straightforward standardized subcontract and purchase order agreements. In addition to assisting the PM with long-form POs, the project engineer and/or field engineer can assist the superintendent with the creation of short-form POs, including cost coding, and potential material pick up from local suppliers, as well as the receipt and verification that the materials ordered and submitted match those that are delivered.

During construction, project engineers, especially senior PEs, also assist the project manager and superintendent in a variety of subcontractor and supplier management activities, including the following:

* Expediting and scheduling subcontractor and supplier deliveries and installations,
* Validating monthly pay requests against work completed,
* Developing early pre-punch lists, as also discussed in Chapter 9,
* Receiving RFIs and submittals, logging, and processing, as discussed in Chapter 11,
* Performing several close-out activities, as discussed in Chapter 14, and/or
* Logging and validation of change order pricing, as will be discussed in Chapter 20.

Summary

Subcontractors and suppliers perform 80–90% of the work on a typical building project, less so on a heavy-civil project. The GC's success depends upon subcontractor success. During the initial work breakdown for a project, the GC's estimator determines which items are to be subcontracted and which of those to include in each subcontract. Based on this subcontract plan, the procurement team crafts a specific scope for each subcontract and purchase order. Once the scope of work has been developed, the project manager utilizes a standard subcontract or purchase order format and prepares specific terms and conditions for each agreement.

Estimators and project managers should pre-qualify subcontractors and suppliers before requesting bids or proposals. Best value subcontractors should be selected based on their ability to provide the greatest overall contribution as members of the project delivery team. After subcontractor and supplier selections, prices are entered on the subcontracts and purchase orders, and the GC's cost control system is updated, as was discussed in Chapter 16. Subcontract agreements will then be executed by the GC's CEO and the vendors' corporate officers.

After buyout and award, the project manager, project engineer, and superintendent work together to manage all the subcontractors and suppliers. Their responsibilities include scheduling supplier delivery and subcontractor installations to ensure that the project is completed on time, within budget, and in conformance with contract quality requirements. The success of the project is dependent on a viable schedule that provides reasonable notice to all subcontractors regarding the scheduled start times for their phases of work. Experienced senior project engineers play a major role in subcontract management, which includes monitoring quality performance, processing RFIs and submittals, issuing change orders, and ensuring prompt payments.

Review questions

1. Why do heavy-civil contractors perform more direct work than typical commercial contractors?
2. Why do commercial general contractors use subcontractors rather than self-performing all the work on a project?
3. Why would a GC (and/or client) choose to pay a vendor for off-site stored materials?
4. What criteria should the superintendent consider when allocating a site area for vendors' materials?
5. What differentiates a subcontractor from a supplier?
6. What is the difference between a request for quotation and a request for proposal?
7. What risks do general contractors incur by using subcontractors?
8. Other than cost, why might a GC require performance and payment bonds from one subcontractor but not from another?
9. Why is subcontractor bid shopping considered unethical behavior?
10. How are mock-ups used for project quality control?
11. Why is a subcontractor's ability to finance their cash flow requirements on a project of concern to the GC?

Exercises

1. Write a clear scope of work for the shoring subcontractor for the Rose Collective mixed-use apartment (MXD) project. Address as many of the contract issues as possible bulleted within the Scope section above.

2. What are five criteria that you suggest be used to pre-qualify mechanical contractors for the Rose MXD project?
3. What basis do you suggest the project manager and senior project engineer use to select the roofing subcontractor for the Rose MXD project?
4. Other than the examples listed in the body of this textbook or in our example figures, name two building materials for our I-90 heavy-civil case study project that would be purchased with a short-form PO and two materials that would be purchased with a long-form PO.
5. By utilizing the Rose MXD project schedule on the companion website and scheduling backward, provide approximate dates of (a) award, (b) submittals, (c) delivery, (d) installation start, and (e) installation completion for one subcontractor and one major material supplier.
6. As a senior project engineer and considering everything necessary to do an adequate job, draft a contract agreement for either of our case study projects for (a) one of our major material items noted above, and/or (b) one off-the-shelf material item, and/or (c) a subcontractor.
7. Prepare an argument for why a pay-when-paid clause is (a) fair or, conversely, (b) unfair.
8. Other than mechanical, electrical, landscaping, and curtain-wall work, what subcontracting scopes might be delivered as design–build on (a) a typical commercial project or (b) a heavy-civil project?

References

ASPE. (2011, May). American Society of Professional Estimators (ASPE) Code of Ethics. Retrieved September 29, 2017, from www.aspenational.org/?page=CodeofEthics

Osmanbhoy, N. M. (2015). Closing the contractual circle: investigating emergent subcontracting approaches (Unpublished master's thesis). University of Washington. Retrieved September 29, 2017, from https://digital.lib.washington.edu/researchworks/handle/1773/33485

20 Managing changes

Introduction

As we have discussed throughout this book, managing built environment projects usually relies on developing plans and implementing them. For instance, owners and designers plan for a designed project scope, and then they hire construction firms to implement the plan. General contractors plan for a timely and cost-efficient delivery of the scope, and then they implement their plans, often with the support of specialty contractors. However, plans are based on assumptions, and it is common for these assumptions to change over a project's life.

Changes can occur at any time during the project development phases, even before a construction firm is selected. For instance, a major retailer may have decided to open a mega-store in a metropolitan area, only to discover that the major city within the area will obstruct permitting to large retailers. Instead of fighting the decision in court, the retailer decides to go around the problem by finding a new location for building the facility beyond the city limits but within the same metropolitan area.

Later on, during procurement, *addenda* can be used to modify the scope of the project before it is bid or the contract is awarded. *Change orders* accomplish the same thing after award. In addition to modifying the scope and terms and conditions of the agreement, a change order often impacts the cost and schedule. Change orders are also known as contract modifications. They may be additive, if they add to the scope of work, or deductive, if they delete work items. However, they do not need to change the scope, because change orders can be used to simply make changes to other contractual terms that affect the project development without adding or deleting scope. As change orders occur on most construction contracts, managing them is an important project management function. In this chapter, an overview of different types of changes, functional and legal, will be provided. Change order management and pricing and the roles of experienced project engineers in the change order process will then be described. Project management applications with respect to change orders will also be expanded upon in Chapter 22.

Types of changes

In this section, project changes will be discussed from two perspectives: (a) functional changes and (b) legal changes.

Functional classification of project changes

Functionally, changes can be classified as changes to project scope and changes to project development. Changes to project scope can be further divided into *required* and *elective* changes (CII 1994).

Required changes to project scope are those that the project team must implement and cannot avoid. These changes are often necessary to (a) "meet the basic, defined venture/business objectives,"

(b) "meet regulatory or legal requirements," or (c) "meet defined safety and engineering standards" (CII 1994; p. 15). In addressing these necessary changes, "the project team must determine if the project remains viable" (CII 1994; p. 15).

For instance, the project plans of a new industrial facility must be changed once the state environmental agency enforces stricter environmental standards that the designed facility would not be able to meet. In this case, the project owner must redesign the facility to meet the standards, or the facility will not be able to operate at capacity. Similarly, a state transportation department must change the geometric design of a road interchange, because it discovers the designed curve would be unsafe to motorists. In this case, the project owner does not have a choice but to defend themselves against future potential lawsuits and, potentially, criminal charges if anyone dies or gets injured on the road.

Elective changes to project scope include all other scope changes. They "are those that are proposed to enhance the project, but are not required to meet the original project objectives. Elective changes, therefore, are those that may or may not be implemented" (CII 1994; p.15). These changes do not need to happen but are made to satisfy the desire of some project team member. For instance, the project owner of the industrial facility above may decide to change the design to accommodate a new industrial process that has recently been demonstrated to be more efficient. In this case, the project owner does not have to implement the change, but the implementation makes for a good business decision. Similarly, the state's department of transportation decides to change the geometric design of another interchange to allow motorists to drive it at a higher maximum speed, which would allow the interchange to accommodate higher traffic volumes. Again, this change is not required, but it would allow the interchange to better accommodate future increases in traffic volumes.

Changes to project development are those that are necessary to achieve the original scope of work and achieve the project objectives (CII 2017). A common example is when specified materials are not available, so the project team is forced to select similar materials, possibly with a longer lead-time and different costs. Other examples include price and currency escalations, changes in labor productivity, and design errors and omissions.

Legal classification of project changes

As project delivery relies highly on contractual relationships, project changes often have legal implications. Legally, changes can be classified into three groups: (1) *directed* changes, (2) *constructive* changes, and (3) *cardinal* changes.

A *directed change* is ordered by the owner and implemented by the contractor. The instruction to implement the change can manifest itself either as a change order directive or as a change order. These documents will be discussed later in this chapter. The process for implementing a directed change is usually described in a change order contractual clause within the contract. This also describes how cost and time implications will be taken into account in the form of corrections to the contracted cost and time. These types of changes need to be directed by the owner before occurring, even if they are initiated by the contractor through a change order proposal or a request for information.

Constructive changes are also known as changes by implication and include those changes that are not ordered by the owner but that either occur or are necessary to achieve the project objectives. Whereas constructive changes are not explicitly requested by the project owner, they may have cost and time implications for the construction team. Sometimes, a contractor can propose a change to the project owner, who can agree to the change by accepting the change order proposal.

Alternatively, contractors may elect to submit a claim for additional time and/or compensation in response to what they qualify as constructive changes. The condition for submitting a claim for a constructive change is that the change is originated by an act or omission of the project owner or their representatives that has the same outcome as a formal change order. Constructive changes clearly do not follow the change order contractual clause. Therefore, these changes are not as intuitive to understand, so it is better to provide some examples.

As described in the mixed-use development (MXD) case study, a developer called Rose Collective LLC has contracted Reliable Construction Company (RCC) as the general contractor for the MXD project in Portland, Oregon. While negotiating the contract for Rose Collective, the project representative, Mr. Warren, has agreed to provide the bathtubs for the apartments that will be installed by RCC and its subcontractors. When the general contractor develops the project schedule, they inform Mr. Warren that the bathtubs should be available for installation at a given day. However, the project owner fails to meet this deadline and provides the bathtubs one month after they were expected. As a result of this change in delivery schedule, the general contractor and their subcontractors need to rearrange the schedule and work around the missing bathtubs, which creates inefficiencies in their workflows. Moreover, once the bathtubs are available, the construction team needs to accelerate the installation to complete the project by the contracted completion date. Given that the inefficiencies and the acceleration produce extra costs for the construction team, RCC decides to submit a cost-compensation claim based on the actions of the project owner and their representative. Even if Rose Collective LLC had agreed to a one-month delayed completion to take the delay into account, the construction team would at least seek compensation for the increase in indirect costs.

As described in the highway overpass case study, a public agency—the Montana Department of Transportation—has contracted Gateway Construction Company (GCC) as the general contractor for the reconstruction of a highway overpass that is wider than the existing overpass. The design assumes the existing foundations will be able to accommodate the piers for the new overpass. After demolishing the existing piers, GCC finds out that the existing foundations were also damaged by the same seismic actions that required the overpass reconstruction. As a result, some immediate temporary structure needs to be built to maintain the stability of the retaining walls while a new foundation that includes piling is designed and built. These actions require postponement of the project contractual completion and result in extra costs to build the temporary support structures and to mobilize pile driving equipment to the site. Therefore, GCC decides to submit a claim for extra time and cost compensation to the state based on the defective design. After reviewing the circumstances surrounding the claim, the project owner decides to interpret the claim as a change order proposal, agrees on the implemented and proposed changes, and decides to adjust the contractual completion and cost.

In both examples, the project owner has not directed the change, but the contractor was forced to take action to compensate for an act or omission by the owner, which explains why construction changes are often introduced as changes by implication.

Whereas directive and constructive changes can be handled while the construction contract is kept in place, addressing a *cardinal change* would require adjustments that are beyond the scope of the contract. This is because they would require the contractor to perform work beyond the types of changes that were expected when the contractor entered into the contract. To address these circumstances, case law has established the *cardinal change doctrine*, which provides a way for contractors to walk away from a contract, while retaining the right for claims. The cardinal change doctrine can only be invoked under circumstances that are rare in the industry and need to be proven, so contractors rarely follow this path and only when they deem it is necessary. As a result, project engineers are rarely involved in managing cardinal changes, except when they may be called to provide testimony during legal proceedings. Such testimony is intended to help

> **Change Order Sources**
> - Designer-generated scope changes, change in material specifications
> - Contractor-generated material or detail change or value engineering proposal
> - Design errors, discrepancies, or lack of coordinated multi-discipline design
> - Discovery of unknown or hidden conditions
> - Municipality-generated changes, code interpretations or permit review changes
> - Not-in-contract equipment discrepancies, utility connections, size variations
> - Client scope changes, additions, or changes to the program
> - Specified product is unavailable or no longer manufactured
> - Interference from third party or another contractor not under the control of the general contractor
> - Unusually adverse weather

Figure 20.1 Change Order Sources

substantiate the nature, impact, and quantity of changes that the cardinal change produced and the dramatic effect on the project delivery.

Change order sources

When changes require a variation of some of the agreed-upon contractual terms, the issuance of a change order is needed. Change orders arise from a variety of sources or causes. Figure 20.1 lists some common sources of change orders. The four most common are elaborated on in this section.

The largest count of change orders, but not necessarily the most expensive individually, usually comes from *design errors and omissions*. These can arise for a variety of reasons, including:

- The designers may not have sufficient funds, resources, or time to do a complete design;
- Project owners will contract with the sub-consultants/designers (such as mechanical or electrical engineers) directly and not through the architect. This may result in multi-disciplinary documents that are not fully coordinated;
- The project is highly complex, so changes are more recurrent to deal with this complexity; and
- Project team members are human, and innocent human error or oversight can always happen.

Project owner requested *scope or programming changes* are among the most expensive change orders. These changes are classified as elective, as discussed above. For example, the Rose case study developer decides the addition of irrigated sidewalk planters with rose bushes of varying colors would be appropriate on this Portland MXD project and asks RCC to prepare a change order proposal (COP) for the addition.

A third category of change orders stem from uncovering *unknown site conditions*, also known as *differing site conditions*, which can be among the most difficult type of change order for a project owner to accept. These generally result from inadequate site exploration prior to starting the design process. Hidden or latent conflicts or conditions are common in site work (e.g., unknown buried debris) or remodeling (e.g., a rotten wood structure) and may use up the project owner's contingency early in the project. When the contractor encounters hidden site conditions that adversely impact construction, the general contractor's (GC's) project team must "promptly" notify the architect and project owner and provide the designer with an opportunity to make an inspection. "Notice" procedures will be discussed later in this chapter.

Equipment or materials supplied by the project owner and not in the contractor's scope, or *not-in-contract*, may result in change orders if the products are not fully coordinated. Some project owners believe that, by contracting directly with subcontractors and suppliers, they save the general contractor's fee. This is common with items such as kitchen equipment, auditorium equipment, furnishings, custom casework, and landscaping on commercial projects or signal equipment and signage for heavy-civil projects. However, these savings come with retaining risks associated with coordinating material and equipment delivery with the contractor's schedule for installation. Problems may surface and the project can be disrupted because of the lack of coordination of these owner-furnished materials. Project owners will often eventually pay much more to resolve these conflicts than they would have paid the general contractor in fees to manage these scope items.

Change order process

As shown in Figure 20.1, contractual changes can be due to multiple sources. Whereas it may vary slightly across projects, the change order process usually follows three stages, as shown in Figure 20.2. Toward the end of this chapter, we will briefly introduce claims, which are often the result of unresolved change order proposals or change directives.

Stage 1: Establish need for a change

If the contractor identifies a differing site condition or some design discrepancy, they will usually start with a request for information (RFI). This may serve as a change notice or trigger a request for a change order proposal by the project owner and their representatives.

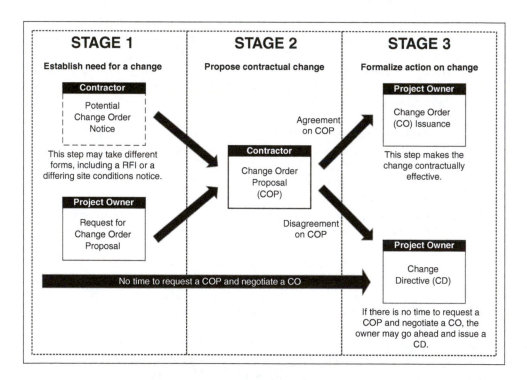

Figure 20.2 Change Order Process

If the project owner wants to initiate a change order (CO), a common route is to inform the contractor of the desired change outcome and request the development of a COP. An owner-initiated COP request provides a description of the proposed change in the scope of work and requests that the contractor provides an estimate of additional or deductive costs. If the proposed change extends the project duration, the contractor will also be requested to quantify the additional time. Although an owner can alternatively issue a change directive (CD), requesting a COP is a softer approach to negotiate a contractual change. However, a CD is sometimes needed if there is not sufficient time to prepare a proposal or if there is a disagreement between the parties.

Stage 2: Propose contractual change

At this stage, the contractor develops a full proposal for a change order. When a COP is produced by the contractor in response to a request by the project owner, it will include information provided by the owner on what will be changed in the scope or development and information on the time and cost changes provided by the contractor.

After gathering all the cost information and backup documentation, the general contractor forwards the COP to the project owner or their representatives. If accepted, this results in the issuance of a contract change order. The contractor's proposal describes the change to the scope of work and provides any proposed adjustments to contract price and time. Figure 20.3 reflects a COP from Gateway Construction for the replacement of unsuitable topsoil for the I-90 overpass case study project. COPs do not modify the construction contract. Only formal contract change orders and construction change directives, which are discussed below, modify the contract.

On a typical building project, subcontractors comprise 80–90% of the work, so they often initiate the majority of the COPs. Well-written RFIs and thorough and timely submittals are methods for raising discrepancies and, therefore, are part of the active quality management process discussed in Chapter 9. The project manager or senior project engineer should research the issue and forward the question or submittal to the lead designer. In the case of Figure 20.3, Gateway had initially submitted an RFI about the potentially unsuitable topsoil, and the state engineer forwarded the RFI to the landscape architect, who responded: "I agree this topsoil is unsuitable. Please submit a change order proposal to replace the unsuitable soil."

After reviewing an RFI, the designer—on behalf of the owner—may realize some design or *differing site condition* issue and request a COP from the contractor (lower branch in process stage 1). However, if the designer does not promptly recognize the issue, then the contractor/subcontractor team can still initiate the process (upper branch in process stage 1). For instance, the subcontractor affected by the issue will notify the GC about cost and schedule impacts from the architect's response. If the GC's team believes a change order is warranted, they will send a COP request to the subcontractor. The project manager or project engineer uses the subcontractor's response to their COP request to prepare the GC's COP, similar to the one illustrated in Figure 20.3. The following guidelines should be used when preparing a change order proposal:

- COPs should be individually numbered,
- A detailed description should be included,
- Direct labor hours, wage rates, and labor burden should be itemized,
- Direct material and equipment costs should be summarized and backed up with detail,
- Subcontract costs should be listed separately from direct costs and itemized or backed up, and
- Markups should include fees, taxes, insurance, contingency, bonds, and sometimes jobsite general conditions.

Gateway Construction Company
2201 First Avenue
Spokane, WA 99205

CHANGE ORDER PROPOSAL

COP Number: 25   Date: 7/1/18

Description of work: On-site material was not suitable for landscape topsoil as had been indicated in the specifications and geotechnical report. This work proceeded under verbal direction of the State's landscape architect.

Referenced documents: Phone memo dated 6/15/18 and RFI 123 in process

COP Estimate Summary:

				Extended Cost:
1	Direct Labor:	10 hours @ $41/hour	(See att'd)	$410
2	Supervision:	2 hours @ $55/hour		$110
3	Labor Burden:	20% of labor:		$105
4	Safety:	2% of labor:		$8
5	Total Labor:			$633
6	Direct Materials and Equipment:	1000 CY @ $20/CY	(See att'd)	$20,000
7	Small Tools	1% of DL		$4
8	Consumables	1% of DL		$4
9	Total Materials and Equipment:			$20,008
10	Subtotal Direct Work (Items 1 through 9):			$20,641
11	Subcontractors (See attached subcontractor quotes):		Landscaper	$5,000
12	Overhead on Direct Work (w/fee) Items:			$0
13	Fee on Direct Work Items:	13% of direct L & M		$2,683
14	Fee on Subcontractors:	13% of subs		$650
15	Subtotal Overhead and Fee:			$3,333
16	Subtotal all costs:			$28,974
17	Liability Insurance:	1% of subtotal		$290
18	Total this COP # 25			**$29,264**

This added work has an impact on the overall project schedule, the extent of which cannot be thoroughly analyzed until after the change is incorporated into the contract and the work has been completed.

Please indicate acceptance by signing and returning one copy to our office within five days of origination.

Approved by:

Jeremy Short   7/6/2018

Montana State Resident Engineer   Date

Figure 20.3 Change Order Proposal

COPs should be presented by the project team in a fashion that makes it easy for the project owner and their representatives to approve. Before initiating any work on the proposal, the general contractor should have the project owner sign their approval. All relevant supporting documents should be attached. Some examples of COP support documents include:

- Originating RFI or submittal,
- Relevant drawings, specifications, and photographs,
- Copies of subcontractor and supplier quotations,
- Quantity take-off (QTO) and direct work pricing recap sheets, and
- Letters, memos, meeting notes, and daily diaries.

Subcontractors should provide the GC with the same level of detailed backup. All COPs should be tracked with a log. As soon as a potential change condition arises, it should be assigned a number and entered into the COP log, similarly to the process for managing the RFI or submittal logs.

If the project owner accepts the COP, the rest of the change order process is relatively straightforward. Often, however, there will be questions related to the request and subsequent negotiations. This is to be expected and is part of working as a team. Remember that the proposal is usually negotiable. The project management team is not taking a hard line approach at this time. A weekly change order proposal meeting, outside of the regular owner–architect–contractor (OAC) construction coordination meeting, is a good way to discuss and resolve change issues. It is beneficial to keep change order discussions out of the coordination meeting, because the discussion of extra costs has a way of undermining the communication process.

Stage 3: Formalize action on change

Change orders and change directives are two separate instruments to implement contractual changes on a project. A *change order* is a formal contract-revising vehicle that is utilized to modify a contract after construction has commenced. On building projects, the change order is generally prepared by the architect. If the designer or project owner and the GC have negotiated a mutually acceptable adjustment in the contract price, schedule, or both, a change order is executed to modify the terms of the contract. The supplemental or special specifications of most prime contracts contain the change order procedures. ConsensusDocs Form 202 and AIA document G701 are common formal change order coversheets and are preferred by many contractors, project owners, and architects.

A change order can be used for one large issue or a group of smaller approved COPs. Some project owners choose to issue monthly formal change orders to incorporate all change order proposals approved during the month. This is a good process for the GC if the CO is executed before the monthly pay request is drafted, because it allows the CO to be invoiced this month. Change order proposals do not modify the terms of the contract and are not added to the schedule of values (SOV) for pay purposes until incorporated into the contract by a formal contract change order. The contract documents should be annotated with all scope changes contained in each change order. Once the change order has been signed by the architect, owner, and contractor, the contract scope, price, and time have been modified. The contractor's signature indicates that they agree that the adjustment in price and time adequately compensates for the added work. After the general contractor's contract has been modified, the project manager or an experienced project engineer will modify the appropriate subcontracts and purchase orders for major material suppliers.

If the project owner and general contractor have agreed to cost and schedule impacts from a COP, the formal contract change order will incorporate the COP into the agreement. If the project owner does not agree to a change order proposal submitted by the GC's project manager, the project manager may submit a claim by using procedures that are introduced later. Alternatively, if the project owner and the GC have not negotiated a mutually agreeable COP, but the project owner still wants to proceed with the change, a change directive is issued. AIA construction change directive (CCD) form G714 is a common formal change directive coversheet. The prime contract may specify that a CD is to be used to allow the work to proceed without an agreed upon change order. A construction change directive may also be used when the change must be implemented before the GC's jobsite or home-office change order estimator has had time to evaluate its cost and time impacts. An alternative to the AIA CCD form is the ConsensusDocs Interim Directed Change Form 203.

A CD is a document that modifies the scope and potentially the cost of a contract. Often the CD indicates that the general contractor is to proceed with the changes and that an appropriate adjustment to contract price and time will be negotiated later. Upon receipt of a CD, the GC determines the impact of the change and submits a change order proposal requesting an adjustment of contract price and time. Also upon receipt of a change directive, the project manager is required to issue a similar direction to affected subcontractors and suppliers. The CD changes the contract scope when it is signed by the architect and project owner, but it does not formally change the contract price or time unless it is also signed by the contractor. If the contractor refuses to sign the change directive, any change in contract price and time will occur only after the architect, project owner, and contractor have negotiated a separate change order. However, the work will still have been accomplished. This may also result in a claim filed by the GC, as discussed later in this chapter.

Pricing change orders

In order for the GC and their subcontractors to be paid for extra work performed, they must have their COPs approved. The easiest way to achieve this goal is to be realistic with respect to pricing on direct and subcontract cost estimates. Overly inflated prices will only delay the process. Quantity measurements are generally verifiable and should not be inflated. Wage rates paid to craft employees are verifiable and should not be inflated. Subcontractor quotes should be passed through as is without adjustment—unless incomplete—from the GC to the project owner. The subcontractors and suppliers should practice the same procedures with their second- and third-tier firms. Labor productivity rates should be derived from preapproved resources. Material prices should be actual and verified with invoices or quotations. Any deviations in the above suggestions may damage team building and trust among the parties and/or delay COP approvals. After all the direct work and subcontractor work is totaled on the COP, the GC will add a series of markups to the estimate. Some of the markups that may be added to the bottom of a COP include:

- Bonds and insurance,
- Cleanup or rubbish removal,
- Consumables and small tools,
- Fees on direct and subcontracted work,
- Jobsite general conditions,
- Foremen and superintendent time,
- Labor burden and taxes,
- Transportation and hoisting,
- Safety, and
- Contingency.

The markups are often cumulative and their total effect on the actual cost of the work can be substantial. Markups, or percentage add-ons, are utilized by some GCs and subcontractors as fee enhancement and recovery for other indirect costs. Project owners often get frustrated with these add-ons. They do not understand why they have to pay more for the change than simply the direct costs. Most of these markups are required, and sometimes it is just the presentation that makes them difficult to accept. An alternative approach is to figure the exact impact for each item, for example, by listing each additional small tool required for this COP and the appropriate unit prices, but this is cumbersome.

An in-depth discussion of each markup is an interesting and worthy endeavor but one warranted for a more advanced construction management textbook. Here we will elaborate on just the fee and general conditions. General contractors usually receive a 3–8% markup for the fee on building projects, and heavy-civil GCs can receive markups in excess of 10% for infrastructure projects. Residential projects and smaller commercial and tenant improvement projects will also see up to 10% in fees. This rate is usually the same percentage fee that was used on the original estimate and is stated or allowed in the contract. Home-office overhead costs are assumed to be included in the fee and are not typically a separate add-on.

Jobsite general conditions are not customarily allowed for building general contractors as a markup, unless they can be substantiated on individual change issues or it can be proven that the scope of the change will extend the project schedule. In order for the jobsite project team to prove that additional general conditions and/or time are warranted, they need to provide documentation to back up the time and costs, such as a very detailed construction schedule. Contractor attempts at recovery for additional jobsite general conditions, time extensions, home-office general conditions, and impact costs are often contentious, especially in the COP process. More often, they will fall into the claim category discussed below.

Subcontractors typically receive a higher fee percentage than the GC, because their volume of work is less and their direct labor percentage and resultant cost risk is higher. Subcontractors may receive a 10% fee and an additional 10% overhead. Both of these rates will depend upon how many of the markup items bulleted above are assumed to be in the fee or overhead or are allowed in addition to the fee.

Contract issues

One of the first steps in the estimating process is for the GC's chief executive officer (CEO) and project manager to carefully study the proposed contract language and the front-end or special conditions of the specifications. Some of the issues that may be discovered in this initial review relating to change orders include:

- *Notice*: Contracts require a strict timeframe from when a contractor realizes a potential change of scope has occurred and when they must (a) notify the designer or project owner of the change and (b) have it priced and presented to the designer or project owner. Terms such as "As Soon As Possible," "Immediately," or "Promptly" are not advisable and are subject to interpretation. Timeframes—such as within 7, 14, or 21 days—are easier to enforce but must be reasonable to be fair to both parties;
- *Differing site conditions*: Contractors are required to research conditions, either on the jobsite or between the separate contract documents, which are potentially different than what was represented. If a contractor proceeds with work that is in conflict with the designer and project owner's intentions, then the contractor is taking the risk of an incorrect interpretation and may be required to perform rework at no increase in cost. As discussed previously, RFIs are the tool

most commonly utilized to request interpretation and provide notice of a changed condition. Timing is relevant here as well; and

- *Response to RFIs*: Some RFI formats dictated by the project owner or designer may require the contractor to present cost and time impacts at the time the RFI is asked. This is difficult for the contractor to do until they have received a response. In addition, once the RFI has been answered, the designer and project owner may assume there are not any cost and schedule impacts unless the contractor responds within a stipulated timeframe, similarly to the notice described above.

It is each contractor's responsibility to study these and other contract clauses and either accept the risks and proceed with the bidding process, increase fees and contingencies, or pass on the project and pursue another.

Claims

Claims can be defined as written requests by one of the contractual parties for extra time or additional compensations as a result of actions or omissions by another party during the project delivery. Claims are considered by many to be the adversarial cousin to change order proposals or contract change orders. Claims often stem from a COP that has been rejected by the project owner or a CD that was forced on the contractor. If no limitations are provided in the contract, there is the risk of many claims arising late in the construction process when contractors have realized that they did not achieve their anticipated estimate or schedule. To address this issue, project owners usually incorporate timeframes for when a claim must be filed in the contract, for example, within 21 days of occurrence. The value of a claim is also customarily much larger than a COP. The COP may be valued at $30,000 on our example project for a discrepant site condition, whereas a contractor may file another separate claim for a different cause at substantially higher values, such as $500,000, $1,000,000, or more because of a time extension beyond their control.

The path for claim resolution is also dictated in the contract. The least expensive and most expedient solution is *prevention*, which requires all parties to be diligent about their communications and resolve issues at the lowest level possible (i.e., the jobsite) before they have escalated back to the home office. If a conflict arises, the parties are encouraged to *negotiate* the solution, also at the jobsite level. The COP and even the CD processes described above are the preferred, less expensive paths. If an issue gets past prevention and negotiation, then the claim resolution process, which is described in the prime agreement, will usually include the following options, in order of least to most investment of cost and time:

Mediation – Arbitration – Litigation

Mediation is an effort to settle the claim through a mediator, who will try to find points of agreement between the parties and make the parties agree on a settlement. *Arbitration* is another approach for resolving disputes outside of court, through an arbitrator or an arbitration panel, who act like a judge or a panel of judges in making a determination on the dispute. Last, court *litigation* is the traditional legal approach for resolving disputes. More recently, some contracts have been appointing a three-member *dispute resolution board* (DRB) at the beginning of projects; this board will advise the parties on potential dispute resolutions before they have escalated to attorney involvement. The cost of a DRB is likely more than mediation but less than arbitration. The concept of disputes, claims, and claims resolution is definitely an advanced topic. Only experienced PEs will be asked to support the PM and project executives in these functions, as addressed in Chapter 22.

Project engineering applications

Senior or advanced project engineers may take a more project managerial role with respect to change orders than with RFIs or submittals. The project engineer (PE) can assist the project manager and superintendent in a variety of fashions with respect to change orders, including:

- Identifying RFIs that could potentially result in change order proposals,
- Soliciting subcontractor and supplier input on all potential COPs,
- Performing QTO and pricing of direct work portions of COPs,
- Gathering the backup documentation essential for submitting a thorough COP,
- Maintaining the COP log,
- Drafting subcontract modifications after CO approval, and/or
- Supporting the project manager (PM) with research and backup for contract change orders and potential claims.

Summary

Changes happen daily in the construction industry and on just about every project. It is up to the whole project team to deal with them in the most expedient and fair manner possible. Change orders and change directives arise from a variety of sources, including differing site conditions, scope additions, and discrepant design documents. Change order proposals may be requested by the project owner or initiated by the general contractor. An owner-initiated COP request describes the proposed change in the scope of work or development and instructs the construction team to provide an estimate of the cost and schedule impact. A GC- or subcontractor-initiated proposal describes the proposed change in the scope of work and requests an acceptable adjustment to the contract price and/or time. The project team maintains a change order proposal log to track all change order proposals, whether generated by the project owner and its representatives, the GC, or subcontractors. Once the GC and the project owner have negotiated a mutually agreeable adjustment in contract price and time, a formal change order is executed to modify the contract. If the project owner wants to proceed with a change but has not negotiated an appropriate adjustment to the contract price and time, a change directive is issued. This allows the work to proceed and the exact cost and schedule impacts to be determined later. A claim differs from a COP or contract change order in that it is a request for time and/or cost from the contractor that the project owner has not yet agreed to accept. Claims are ultimately resolved according to the dispute resolution process, as described and agreed to in the prime contract.

Review questions

1. What is a change order?
2. What are four typical causes of change orders?
3. How does the use of owner-furnished materials result in change orders?
4. How do designer responses to RFIs result in change orders?
5. What factors should a project manager or advanced project engineer consider when analyzing the impact of a proposed change order?
6. What recourse does a contractor have if the project owner denies a COP?
7. How does a COP become a formal contract change order?
8. What is the difference between a change order and a change directive?
9. What action should the jobsite management team take upon receipt of a change directive?
10. Why might a project owner decide to issue a change directive?

Exercises

1. Prepare a COP with a value greater than $20,000 related to one of our case study projects. Include a COP cover that allows the project owner to sign and approve. Include all relevant supporting documents, such as quantity sheets, recap sheets, subcontractor quotes, sketches, RFIs, submittals, and/or memorandums. Sell the change!
2. Prepare a formal contract change order by utilizing an AIA or ConsensusDocs coversheet and incorporating the change order proposal prepared in Exercise 1. Note the schedule impact, if any.
3. Assume you are preparing a COP that has estimated costs of $10,000, $15,000, and $45,000 for direct labor, direct materials, and subcontractors, respectively. Apply reasonable markup values for as many of the potential markups bulleted in this chapter as you can. What is the total COP value, and what was the cumulative effect of these markups?

References

Construction Industry Institute (CII). 1994. Project change management. Special Publication 43-1. Austin, TX, USA: Construction Industry Institute, the University of Texas at Austin.

CII. (2017). Construction Industry Institute (CII) Glossary. Retrieved September 29, 2017, from www.construction-institute.org/resources/knowledgebase/about-the-knowledge-base/glossary

21 Managing project risks

Introduction

This book has so far discussed various project engineering concepts, including what constitutes project success, as well as techniques for planning and controlling for a successful project outcome. The rigor required to identify project success—to plan for it and to control how the plan is working—is a requirement for survival in construction, but it may be fruitless if things do not go according to the plan. In the construction industry, individuals and companies expect to be exposed to risks of different magnitudes; some risks will affect a task, others could affect the whole project, and some may affect a company's survival or an individual's reputation. The ability to forecast what risks may arise on a project and how to mitigate them is paramount in establishing the right conditions to successfully deliver a project and succeed as a professional or as a company.

For example, a heavy-civil contractor hired to build a major highway overpass must select the appropriate crane system to handle the expected loads. If the selected crane system fails under load, fatalities may occur, as well as substantial delays in project completion. The contractor will immediately be hit by the cost of repairs to the structure and replacing the crane system. Later, they may be hit by liquidated damages and see an increase in worker compensation insurance rates that will make future bids less competitive. Lastly, workforce productivity will take a hit as a result of the low morale because of the fatalities of fellow coworkers.

Although risk management is a duty for all parties involved in a project, including the owner and the design firm, project managers and superintendents are those who manage construction risks on a daily basis. However, experienced construction project engineers may also support site management with risk management duties. This chapter provides an overview of basic concepts and practices for construction risk management. Risk management is an advanced construction management topic, but it is one that all team members must be aware of, including project engineers.

Risk management process

A diligent approach to risk management should incorporate four major elements that are implemented on a project throughout its delivery. These elements and their sequence are represented in the risk management cyclical process shown in Figure 21.1 and are described with examples in the rest of this section.

Risk identification

The first step of the risk management process is to recognize risks, which means the project team should identify events that can potentially affect either the project as a whole or one of the project

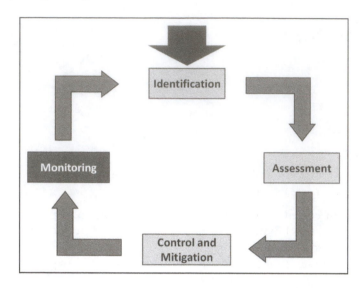

Figure 21.1 Risk Management Cycle

team members, including the project owner, the designer, the general contractor, or any of the subcontractors and vendors. Examples of risks that could affect a project as a whole could be the discovery of underground conditions that make it impossible to build the facility as designed or the occurrence of a catastrophic weather event that damages the structures under construction. Examples of risks that could affect individual team members and the project delivery include the sudden passing of the project superintendent, which would affect field leadership, or a fire destroying the facility of the steel fabricator selected for long-lead structural elements.

Risk identification is a crucial step, because risks can be classified into known and unknown categories. Project team members may be able to mitigate risks only if they are known, but what is unknown cannot be planned for. Not knowing a risk almost equals accepting that risk, because it is impossible to select any risk control or mitigation approach for something that is unknown. Thus, unknown risks are the worst type of project risks. Although it is impossible to forecast all unknown risks, it is certainly a shortcoming if the unknowns include risks that should have been recognized by the individuals who have been tasked with managing risks on the project.

To reduce the chance of accepting unknown risks, project teams often utilize various tools and undertake risk identification exercises. The use of *risk checklists* is quite common among organizations in the built environment. Whereas large project owners and contractors develop and maintain their own checklists, various organizations have developed lists for use by anyone. For instance, the Construction Industry Institute (CII) performed a study that identified 107 risks that are commonly encountered on construction projects (CII 2006).

Risk identification sessions are also performed by project parties, individually or collectively. Risk identification sessions are better performed by a diverse group of individuals within the project team who may bring different perspectives and reduce the occurrence of absorbing unknown risks. For instance, it was mentioned in Chapter 17 that any contractor evaluating a bid on a project will have their senior estimator evaluate its exposure by developing a conceptual estimate, even before investing substantial efforts in the estimating process. As part of this initial analysis, the contractor's executive management will also evaluate any risks associated with winning the bid, such as the risk of the project owner being financially insolvent; ideally, the contractor's operation manager will also evaluate if the project requires unusual construction means and methods that could carry technical risks.

Risk assessment

The second step aims at evaluating all of the identified risks to assess the range of possible risk events and their potential impacts on the project. Impacts are often assessed along the four pillars to achieve project success; therefore, the impacts of a risk event are assessed in regards to cost, time, quality, and safety. However, it is rare for a risk event to have impacts on all four dimensions.

Contractors perform risk assessments by evaluating (a) the chance each risk may occur and (b) the impacts of its occurrence on the project. The exposure for each risk is directly proportional to how each risk fares against these two dimensions. Therefore, the higher the chance for a risk to occur and the higher the severity of its impact, the higher the risk of exposure. This assessment is usually done qualitatively. However, some larger contractors have developed custom software programs and spreadsheets based on historical data that allow them to apply probabilistic risk analysis methods. To this end, Monte Carlo simulations are frequently used to assign dollar values to risks during the procurement phase so that these can be incorporated into contingencies. When risk chances and impacts can be quantified, a risk exposure factor can be computed as the product of the two. However, quantitative risk management is rarely adopted, because its effectiveness relies on accurate mathematical models and data sets that are not always available.

Risk control and mitigation

The results of risk assessment allow the construction team to recommend the most appropriate method for risk control and mitigation. Mitigation strategies can be categorized into four groups:

Risk avoidance

When a risk event is considered probable and of high impact, contractors may avoid it by changing the construction plan to avoid a technical risk. This approach is commonly selected when the risk assessment returns high chances for a risk to occur that would have severe impacts on the project. For instance, a contractor may decide to not bid on a large project if they find out that the project owner lacks appropriate management of the financial cash flow, which would create financial stress on all project participants. Similarly, a heavy-civil contractor competing for a design-build contract for a bridge replacement discovers, while visiting the site, that there are high chances of finding large preexisting concrete structures during the excavation for the footing that would delay the construction completion. However, the contract does not provide relief for this delay. Based on this information, the contractor may now decide to submit a proposal that change the design of the footing to avoid these underground conditions.

Risk retention

Two types of retention exist. First, *active retention* occurs when a contractor evaluates a specific risk event and consciously decides to take it without mitigating it. This means the contractor is accepting the risk and retaining its consequences. This decision results in estimating the cost and time impacts so that the price and schedule can be adjusted accordingly. For instance, a contractor who is pricing a year-long project for a high-rise building in the downtown core of a metropolitan area realizes that deliveries will not be allowed between Thanksgiving and New Year during normal business hours. This contractor may decide to accept this risk, adjust the delivery schedule to avoid deliveries during this period, and revise the price proposal to include overtime for jobsite personnel taking deliveries after business hours.

However, another scenario is called *passive retention*. This means the contractor is retaining a risk because they do not know about it, so they cannot do anything to select any of the other mitigation approaches. As previously discussed, this is the most damaging outcome from unknown risks. For instance, if the above contractor did not know about the city's no-delivery policy, they would end up unknowingly taking this risk. Whereas the contractor could probably later implement some mitigation measures, such as taking deliveries outside standard business hours, their price proposal would not incorporate overtime.

Loss control

Two types of loss control exist, based on either loss prevention or loss reduction. A contractor can elect to implement a *loss prevention* program by reducing the chance of a risk event occurring. For instance, a few weeks into a project, the general contractor realizes that hand tools are missing and probably being stolen by its crew. To prevent these issues from recurring, the contractor can implement a loss prevention program where workers need to take individual responsibility for tools that they check out from the jobsite tool shop.

A contractor can also elect to implement a *loss reduction* program by reducing the severity of the risk occurrence. A typical example is when a contractor raises some of the minimum requirements for safety to reduce the number of events that result in deaths, injuries, or lost time from work. For instance, several contractors have mandatory daily stretch-and-flex exercises, with the intent of reducing the occurrence of musculoskeletal (soft tissue) injuries, such as strains, sprains, and repetitive motion injuries.

Transferring/sharing risks

This is one of the most common methods for mitigating construction risks and mostly relies on shifting risks from one party to another through legal documents. Several approaches exist for transferring risks to other companies or sharing them. These approaches will be discussed in a later section of this chapter.

Risk monitoring

Once a set of mitigation strategies has been selected and implemented on a project, it is the role of the project manager and superintendent to continue monitoring how they are working, as well as the occurrence of unknown risks. Most of the monitoring is embedded into the standard project control and monitoring practices that have been described in earlier chapters. However, some contractors implement procedures for the construction team to periodically meet and/or present to executive management on the status of the project, including its risks and adopted mitigations. These risk management sessions offer contractors an opportunity for receiving feedback from other professionals who are not directly involved in the project. They are expected to help contractors think outside the box, in addition to preventing the occurrence of unknowns.

Transferring or sharing risks

The transfer or sharing of risks with other organizations is one of the most common methods for mitigating construction risks, and it mostly relies on shifting risks from one party to another through legal documents. Several approaches exist for transferring risks to other companies or sharing them. Some of these approaches are discussed in this section.

Contract surety bonds

A common approach for transferring risks is to request *contract surety bonds* for construction firms. As explained in Chapter 3 and diagrammed in Figure 3.2, surety bonds establish a three-party relationship where the construction surety accepts to take certain risks from the principal—usually the construction firm—and to compensate the obligee—usually the project owner—in exchange for a fee. Once requested to provide a bond on a project, the construction surety investigates the project and its participants to evaluate if the chances of certain risk events are low enough to issue a bond. Three types of bonds are prevalent in the industry to cover specific risks: bid bonds, performance bonds, and payment bonds.

For instance, by issuing a *bid bond* on behalf of a contractor, a surety is assuring the project owner that the contractor is submitting their price proposal in good faith, the contractor will enter into a contract for services at the proposed price, and the surety will provide performance and payment bonds if they are needed. If the contractor submits a bid and is the low bidder but then refuses to sign the contract, the surety would compensate the project owner. The amount of compensation would be stated in the procurement documents and is often equal to the difference between the low bid and the second bidder, but it has a limit, such as 5% of the bid price.

Similarly, by issuing a *performance bond*, a surety is assuring the project owner of protection from economic losses in case the contractor is not able or willing to fulfill the contractual expectations. Protection usually manifests in the form of monetary compensation; however, there are cases where the surety decides to provide resources to help the original contractor complete the project or to hire another contractor.

Lastly, by issuing a *payment bond*, the surety is assuring the project owner that the contractor will pay subcontractors, direct labor, and material suppliers on the project. If the contractor does not pay its creditors, the surety will compensate them, which keeps them from filing a lien on the project property.

Insurance

Differently from surety bonds, a contractor securing an *insurance policy* creates a two-party relationship where the insurance company will take certain risks from the contractor in exchange for the payment of a premium. Insurance companies rely upon actuarial analysis to evaluate the chances that a risk event may occur and its financial impact on a project. They then establish a premium price for them taking on that risk. Risks are usually grouped by category and packaged into standard insurance company policies. An insurance policy is a contractual document between the insurance company, the insurer, and the entity or individual that requests coverage against the risks (i.e., the insured). Examples of standard insurance policies for the built environment include professional liability insurance, errors and omission insurance, general liability insurance, property insurance, builder's risk insurance, business owner insurance, worker compensation insurance, and construction vehicle insurance.

For example, a general contractor working on a new hotel in a coastal area may purchase builder's risk insurance with additional coverage for damages from hurricanes, typhoons, and high wind. The policy has a $5,000 *deductible* for damages to installed material, but it does not cover loose material. The construction is nearly complete when the hotel is hit by a mild typhoon that results in $50,000 damages to the exterior cladding, as well as the loss of $20,000 in materials stored on the ground. The insurance company accepts its obligation to pay up to $45,000 in repairs for the damages to the cladding but declines to replace the damaged materials, because the policy placed the responsibility to secure loose material on the contractor.

Contracting and subcontracting

The most common risk management approach is *contracting/subcontracting*, with the contract being used as a risk shifting tool between the contractual parties.

For example, a project owner could deliver a new facility by multi-prime contracting, which means the project owner would be responsible for selecting and hiring specialty contractors and coordinating their efforts. This would expose the owner to the full risk of failing in these duties. Instead, the same owner could decide to reduce and share these risks by paying a fee and hiring a general contractor to manage the construction process and coordinate the specialty contractors. This allows the project owner to use the contract to transfer the risks associated with some responsibilities to the general contractor and to share others.

As a different example, a general contractor (GC) that only performs building projects in a certain geographic region enters in a contract to build a new recycling facility outside that region. Because this GC owns a heavy-civil branch out-of-state, they may decide to self-perform or subcontract out the earthwork for the project. Self-performing the earthwork would mean bringing their own equipment from out-of-state and trying to convince some of their qualified equipment operators to relocate. These prerequisites carry costs and uncertainties, which the general contractor could avoid by instead securing a subcontract with a local earthwork contractor. Subcontracting of the earthwork would transfer the mobilization and technical risks associated with the earthmoving to the subcontractor. However, just hiring an earthmoving contractor does not transfer all of the risks, because hiring the wrong subcontractor, who could fail in their duties, means the general contractor would still be responsible for completing the task. Therefore, risk mitigation through subcontracting is better performed through meaningful subcontractor selection, as was discussed in Chapter 19.

Overall, the golden rule for equitably allocating risks among contractual parties would be to allocate "each particular risk to the party that is best equipped to handle it" (CII, 2006; p.v), which sometimes means sharing responsibilities instead of fully transferring them. Because contracting and subcontracting are some of the most adopted approaches for mitigating risks, the previously mentioned CII study that identified the most common 107 construction risks also identified 14 "hot button" risks that are often inappropriately allocated through contracting and should be carefully analyzed during contract negotiation. Table 21.1 lists and describes these hot button risks.

Joint ventures

Another approach for a contractor to share risks is to create a project-specific joint venture with other contractors. A joint venture is a business entity created for a specific project to pool resources without requiring individual investments. Therefore, an individual contractor with adequate bonding capacity could pursue a large project alone, but this approach may force them to acquire additional human or physical resources that the contractor may not have. This could create a strain on the contractor's finances. As an alternative, this contractor could find another contractor who could contribute the resources that the first contractor lacks, and they could form a joint venture for the project. This approach would allow the project risks to be spread among the members of the joint venture, so that no member would be completely exposed alone. Joint ventures are also formed to pool bonding capacity. Therefore, if a contractor lacks adequate bonding capacity to pursue a large project, this contractor could find partners to pool enough bonding capacity to pursue the project opportunity.

Table 21.1 Construction Risks

#	Risk	Description
1	No damages for delay	A project may be delayed due to the owner's fault, which is outside of the contractor's control. Some contracts make it difficult for contractors to seek damages for owner-caused delays.
2	Consequential damages	Failing project objectives may result in damages beyond the project. For instance, delay in the completion of a warehouse would result in a loss of revenue or in rental costs for the project owner. Similarly, a contractor may not be responsible for a project failure, but its reputation can be affected by it.
3	Indemnity	On a construction project, it is highly possible that people or property will be damaged during construction.
4	Ambiguous acceptance criteria	Quality is a critical pillar to project success. Unclear quality expectations, including for testing and inspections, are expected to increase disputes between project parties and impede success.
5	New or unfamiliar technology	In the attempt of building the best and most up-to-date facility, it is common for project owners and designers to specify new products or technologies. However, what is new is often risky. It is important for the parties to discuss the risks associated with the adoption of the new technology and to allocate them to the party that can best handle them.
6	Force majeure	Projects can be affected by unpredictable and catastrophic events, such as floods, earthquakes, and tornadoes.
7	Schedule acceleration	Some tasks may be delayed that would force the performance of later tasks to be accelerated. Acceleration is conducive to inefficiencies and extra costs in the form of overtime and extra labor. The contract should discuss this risk and the possibility for the contractor to receive compensation when the delay is not of its responsibility.
8	Cumulative impact of change orders	Similarly to schedule acceleration, the occurrence of many changes on a project produces a loss in productivity and inefficiencies. The contractual approach for recognizing extra costs and extra time is important for mitigating this risk.
9	Owner-mandated subcontractors	Selecting the right subcontractor is paramount for a general contractor's ability to succeed. However, there are cases where project owners prefer certain specialty contractors to handle some scope. The contract should discuss who should retain the risk of these owner-mandated subcontractors.
10	Insurance allocation	Certain risks are better transferred to an insurance carrier. The contract should clarify which contractual party is responsible for obtaining insurance and paying deductibles.
11	Differing site conditions	Independently from the survey work done in advance, once construction commences there are chances of finding something different from what was expected. Risks associated to these unknowns should be allocated in the contract.
12	Design responsibility	Design is often performed by multiple parties, including specialty contractors and consultants. Risk of deficient design should be allocated through language in the prime contract and subcontracts.
13	Waiver of claims	The timing and circumstances for obtaining a waiver of claims are conducive to potential risk for contractors. These details should be reviewed while negotiating a contract to ensure that they are reasonable and satisfy the parties.
14	Standard of care	The success of a built environment project relies on the professionals involved. Oftentimes, contracts adopt vague language to require the "highest and best industry standards" or "workmanlike manners." These expressions are ambiguous and should be replaced when possible by unambiguous terms and possibly measurable benchmarks.

Adapted from (CII 2006)

Project engineering applications

The roles and responsibilities of construction project engineers also include risk management duties. Project control processes often include the monitoring of major risks to be performed by all team members, including project engineers. Senior project engineers are also often tasked with supporting project managers and superintendents in evaluating the status of known risks and the measures to mitigate previously unknown risks. If the contractor implements periodic reporting and presentations to executive management on risks, a senior project engineer will be tasked to help the project manager prepare this presentation. Whereas construction risk management tasks are based on the experience and expertise of the project team and do not require the use of specific technology, when quantitative probability risk analyses are performed, project teams may use in-house software or generic commercial software programs. If a quantitative risk analysis program is used, it is probable that a project engineer will be trained to support the use by the project team. Examples of commercial programs include Oracle's Crystal Ball and Palisade's @Risk. These programs incorporate a rich portfolio of tools and analyses that can be handpicked for use on a specific project.

Summary

The construction industry is known to follow cyclical market trends, with ups and downs that expose project team members to various risks. In addition, the delivery of construction projects exposes contractors, owners, and designers to multiple project-specific risks that may affect project success and completion, as well as the reputation and survival of the project. Although risk events affecting task performance are common and often manageable, risk events that affect a project can hit a company's reputation and impede project success.

Risk management is an advanced process to support teams in their efforts to identify, assess, control, and mitigate risks on their projects. The process incorporates four major elements that are cyclically implemented throughout the project delivery. The initial steps of risk identification and risk assessment are crucial to success and build the foundations for later decision-making. Once risks are forecasted and evaluated, the project team will be able to select the most appropriate strategy to control the risks and mitigate their impact. However, for the process to be completely implemented, the project team should continue monitoring how the selected strategies are working and if any unknown risks are occurring. Typical construction risk control and mitigation strategies include the use of surety bonds and insurance, contracting and subcontracting, and joint ventures. Project engineers are actively involved in monitoring major risks. Moreover, senior project engineers may also need to support project managers and superintendents in evaluating the status of known risks and the measures to mitigate previously unknown risks.

Review questions

1. How are active and passive retention different from each other?
2. What are the two main components of risk assessment?
3. A contractor is worried about potential productivity loss from weather conditions on a project in Alaska. To mitigate this risk, the superintendent decides to oversize the direct workforce by 10%. What type of risk control method is the superintendent using?
4. A heavy-civil contractor is asked to bid on a project and enter into a lump-sum contract. The level of design detail for the earthwork is incomplete, so the contractor is forced to roughly estimate the excavation volumes. The contract will not allow additional recovery for extra quantities. What type of risk control method should the GC use?

5. A project manager working for a GC is selecting an electrical contractor for a high-rise building. Three electrical contractors submit similar bids. One of these firms is unknown to the GC. Another firm has developed a great relationship with the GC on previous projects, but it looks like all its best crews and site managers are already assigned to other projects. The third firm has worked well with the GC, and after completing another project they are currently working on, they will transfer those crews to the current project. The project manager selects the third firm. What type of risk control method is the project manager using? Against what risk?

Exercises

1. You are working for an interior finish contractor. A general contractor with a shoddy record and a questionable financial outlook proposes to your company to enter in an exclusive lucrative contract for the next two years. What would happen if the general contractor is not able to secure funds to pay for the interior finishing work? Can you assess the probability for this financial shortfall? What mitigation approach would you recommend?
2. Review the following case examples, then write a short paper to identify and assess the risks and suggest what risk control method the contractor could use.
   a. You are working for a utility contractor that has been hired to build a pipeline through a Native American reservation. The utility informs you that your firm can proceed with construction as soon as permits are secured from state and local governments.
   b. You are working with a major international heavy-civil contractor that is part of a joint venture with a local contractor to build a major tunnel under a metropolitan downtown area. The size of the tunnel is unique, and a foreign corporation is selected as the vendor that will design and build the tunnel boring machine.

Reference

CII. (2006). Project change management. Research Summary 210-1. Austin, TX, USA: Construction Industry Institute, the University of Texas at Austin.

22 Introduction to construction project management

Introduction

There are many good textbooks on the subject of construction project management, but few, if any, that focus on the role of project engineers; hence the reason for this book. Many concepts discussed in this *Introduction to Construction Project Engineering* book are also construction project management topics. This book has provided an introduction to those concepts and discussed how they are applied by project engineers. A project manager (PM) developing his or her leadership skills will mentor a project engineer (PE) by delegating these topics and allowing the engineer to realize success while still retaining overall responsibility for shortcomings. All these project engineering topics may still be performed by the project manager, especially on smaller projects or with smaller general contractors and subcontractors. As indicated, project engineers and senior project engineers report to project managers, and project managers, in turn, report to either a senior project manager or to an officer-in-charge (OIC) or chief executive officer (CEO) with smaller firms. Some project engineers perform many of the same project engineering functions in the field and often report to the superintendent and/or support the superintendent with documentation efforts. Similarly to the organization chart introduced in Figure 3.3, the relationships between the PM, the Superintendent, and the PEs are shown in Figure 22.1.

In this concluding chapter, we will briefly reintroduce many of the same topics but with a focus on project management duties. The latest edition of *Management of Construction Projects, A Constructor's Perspective* (Schaufelberger and Holm 2017) is a comprehensive project management textbook, one that includes many of the project management topics that we introduce here. Many of the subsections in this chapter are brief introductions to full chapters from, and follow the relative order of, that textbook.

Procurement and construction contracts

Larger general contractors (GCs) will have a full-time marketing specialist who calls on potential clients and design professionals and brings projects into the office, either to bid or to prepare a negotiated proposal for consideration. Project managers who have experience in a particular construction market or location, such as our case study example with a mid-rise mixed-use building in downtown Portland, will use their architectural relationships and past clients to procure construction projects in the following order, strongest to weakest:

- Negotiate a project with a client with no competition,
- Receive a request for proposal (RFP) for a negotiated project with a short list of similar GCs, say three firms,
- Receive a request for quotation (RFQ) for a bid project with a short list of similar GCs, say five firms,

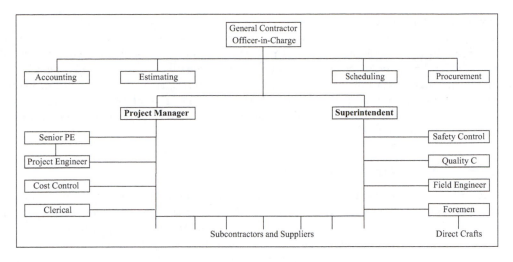

Figure 22.1 General Contractor's Organization Chart

- Attempt a bid at a long-list RFQ that requires some pre-qualification process, or
- Open-bid a project.

Many successful project managers can manage the work and utilize all the construction management tools presented in this textbook, but the PM who can also secure new work will separate themselves from others. In a slow market, where work is hard to come by, he or she makes themselves truly valuable to the construction company. Once the project is landed, the PM will want to be involved in the pre-construction process and take the lead on early estimating and scheduling activities, as discussed below.

In Chapter 5, many different contract agreements were presented. For any contractor, it is beneficial to have an agreement that represents their type of firm and one whose language is, at least, neutral to all parties, if not in favor of the contractor position. Many construction companies are set up so that either a contracting officer or the CEO drafts the prime agreement with the client. An experienced PM may also lead on drafting contracts. On a negotiated project, if the client and architect are neutral to format, the PM can recommend one. The PM can begin the drafting process by inserting his or her language first and then attaching their proposed contract exhibits. On bid projects, such as a heavy-civil unit price road reconstruction, the contract itself is likely included in the project manual or referenced in the instructions to bidders. The contract is the most important of all the construction documents that we have discussed here. All other documents relevant to the relationship and understanding of the parties need to be referenced and included in the contract. An experienced PM wants complete autonomy and control with respect to bringing work in the door, as well as negotiating the contract, if at all possible.

Estimating and scheduling

Similarly to the contract discussion above, the PM will want to take point on developing the estimate and schedule, possibly functioning as bid captain or lead estimator. In this case, the PM will assemble an estimating team that includes his or her project engineers, the superintendent, and others from the office who may have some experience with this type of project. If the construction company utilizes staff specialists for estimating and scheduling, as shown in Figure 22.1, then the PM and superintendent will actively participate in providing construction means and methods and planning and pricing input. Successful project managers and superintendents do not sit in the jobsite trailer drinking coffee and waiting for others to deliver estimates and schedules. Cash flow curves and manpower curves are products of these efforts, as well as cost control planning, work packages, and the jobsite layout plan. Whether the PM or chief estimator takes point, the contractor's CEO will often still make a decision on bid day, after considering the risks involved, and will select the final bid price and fee.

Subcontractor and supplier management

Even more so than with procurement and development of the prime agreement with the client, the project manager plays a significant role in subcontractor and supplier buyout and management. Similarly to estimating and scheduling, some larger contractors also have procurement specialists who will choose subcontractors and draft their contracts. PMs and superintendents should participate actively in this early phase of the construction project, because it is they who will ultimately have to manage the work and guarantee to their superiors and the client that these second-tier firms will perform. The PM wants to hire the best value subcontractors, who not only are cost competitive but also have a record of good performance on schedule, safety, and quality.

Contractors should use standard subcontract and purchase order agreements. Many of these can come from popular sources such as the AIA or ConsensusDocs, but many larger GCs will have had their legal advisors develop formats that are specific to this contractor. The PM will take point on drafting the agreements, especially the larger and riskier ones, such as mechanical, electrical, and elevator subcontracts, as well as steel and concrete purchase orders. The PM should look to the superintendent to also attend subcontractor pre-award meetings and to provide delivery and sequencing input to be referenced in these contracts. PMs can mentor their project engineers at this phase as well, by allowing them to attend buyout interviews and draft many of the less risky agreements.

Progress payments

One of the most important project management functions is to ensure payment for the work performed. A project manager may have all the tools necessary to earn a profit on a job, but if the client does not pay for the work, the contractor will not be able to realize a profit. Some project managers do not acknowledge the importance of preparing prompt payment requests. This is especially true with many subcontractors. If a payment request is not submitted on time, the project manager will likely not get paid on time. Cash management is essential; otherwise, the general contractor may find that they are unable to pay suppliers, craftsmen, or subcontractors. Good cash management skills, just like good communications skills, are essential if one is to be an effective project manager.

Receipt of timely payments is one of the most important responsibilities of the project manager. The exact format for submitting payment requests will vary depending on the type of contract. A schedule of values (SOV) is used to support payment applications on lump-sum, unit price, and cost plus contracts. The project manager is responsible for developing the payment request, making sure payment is received, and subsequently seeing that the subcontractors and suppliers are paid. If payment has not been received on time, the project manager should contact the client to determine the cause. The financial relationship with the client is the project manager's responsibility.

The same scenario holds true with respect to subcontractors and suppliers. The project manager is responsible for ensuring that they are paid promptly. Clients may withhold a portion of each payment to ensure timely completion of the project. This is known as retention. The retention rate is specified in the contract. Liens can be placed on a project if subcontractors or suppliers are not paid for their labor or materials. To defend against liens, clients require lien releases with payment applications. Although the project engineer may become involved in collecting lien releases from the subcontractors, it is the PM's job to make sure that no liens are filed against the client's property.

The project manager is the ultimate accountable party for assembling the pay request, selling it, and receiving payment from the client. This is best achieved through timely pay request processes that abide to the contract requirements, without inflated costs or excessive front loading of the SOV. The PM is also responsible for ensuring prompt payment to his or her subcontractors.

Project controls

Throughout this textbook, we have emphasized the role of the project engineer with respect to project control management. Some project management publications have defined project controls as pertaining to *cost*, *schedule*, and *quality*. This has resulted in the assumption that project success only depends on these three factors. However, project success in construction cannot be achieved if proper focus is not given to *safety*. Figure 22.2 was previously introduced in Chapter 2 as Figure 2.1, and it illustrates the four pillars of project success that should be used to shape processes for project controls. Although many of the project control activities may be delegated to the PE, the PM and superintendent ultimately remain responsible to their client and the contractor's home office to ensure success.

Similarly to other project controls, all members of the jobsite team play an integral role in *cost control*. The project manager will collect actual cost data from the job cost history report and combine that with foremen work packages, often coordinated by the PE, and develop the monthly forecast. Three-week *schedule* development and logistics coordination is the responsibility of the superintendent. The PM will review these schedules, including those developed by the subcontractors, to see if there are any trends that warrant additional focus. If the contract schedule requires updating or revision, beyond just a weekly status report by the superintendent at the

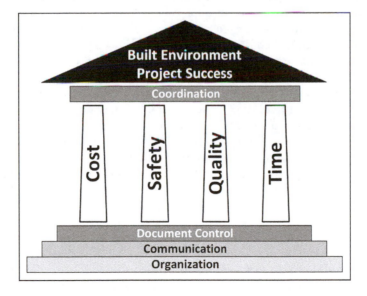

Figure 22.2 Evaluating Project Success

owner–architect–contractor (OAC) meeting, the PM will lead this effort. If there is a new schedule revision, it will need to be included in a change order to the client by the PM and be negotiated with all the subcontractors.

The *quality control* and *safety control* efforts are spearheaded by the superintendent. The PE plays a support role, but the PM will ultimately need to become involved if there are consistent subcontractor violations, reporting back to the GC's home office or the client as needed. Even though the superintendent may lead on "control," the client will still look to the project manager for contract compliance. As previously stated, construction managers do not utilize hammers or saws. Rather the tools in our toolboxes, or more likely computers, are "documents." The *document control process* covers all the topics we have discussed in this textbook, not just requests for information (RFIs) and submittals. Although the project engineer will be actively involved in construction document creation and flow, the PM is responsible, especially in the eyes of the client and the designer, for the accuracy and timing.

Change orders and claims

More so than many of the book's topics, change order management is one area where the project manager will always play a leading role. Project engineers may be requested to gather backup information, but the assembly and negotiation of change order proposals (COPs) with the client is the project manager's responsibility. The ultimate financial success of the project may rely on the management of the change process, especially on lump sum public bid projects. The PM will collect all the approved COPs in a calendar month and have them incorporated into the contract through a formal contract change order. Usually this is targeted before the end of the month, so that approved COPs can be included in the monthly pay request.

Our focus is not on claim development, but rather successful project managers and project management teams who focus on claim prevention. That is what is best for all stakeholders, from the client and designer through subcontractors and foremen. This is achieved through open communication processes and good project management procedures, many of which have been presented in this textbook. If a COP cannot be resolved and a claim is necessary, the PM will likely play a secondary or research role to the home office for assembly, presentation, and resolution.

If a claim cannot be resolved, it must go through a claim or dispute resolution process, which will have been specified in the contract. There are several methods used by the construction industry to resolve disputes. Prevention or avoidance is the least costly and most efficient, and should, of course, be at the top of the list. If a dispute arises, *negotiation* at the jobsite level should be the first technique attempted, because it is relatively inexpensive and not time consuming. If negotiation does not result in resolution, other techniques may be selected in an attempt to find resolution. *Litigation*, or going to court, is usually the last resort, because it is the most expensive and time-consuming. *Mediation, arbitration,* and *dispute resolution boards* (DRBs) are techniques that have been adopted by the construction industry as alternatives to litigation. These three are considered as alternative dispute resolution (ADR) techniques. Dispute resolution techniques are ranked in terms of cost and time impacts, with negotiation being the least costly, litigation being the most costly, and ADR options being in the middle.

Close-out

Unfortunately, many project managers and superintendents demobilize from their projects before final close-out is achieved and the final pay request and retention checks are received. The retention is often approximately equal to the contractor's fee; therefore, PMs should be motivated by an expeditious close-out process. Although close-out is an opportunity for project engineers and foremen to excel and develop project management skills, it is ultimately the responsibility of the PM and the project superintendent to see the project through and gain satisfaction from the client and designer. The best way to achieve an efficient close-out process is for the PM to start it when buying-out the subcontracts by including appropriate language in their agreements. Throughout the course of construction, the close-out log should be maintained, including the preparation of early pre-punch lists. Financial close-out for the PM is critical, both with the client–GC contract and the GC–subcontract agreements. Additional in-house project management close-out activities include as-built estimates and schedules and a lessons-learned report so that all the contractor personnel on this project, and other PMs on other projects, can learn from our successes, as well as our shortcomings.

Introduction to construction project leadership

There has not been a lot written specific to construction leadership, let alone any accepted definition. There are literally hundreds, if not thousands, of books and academic articles written on the topic of leadership, but most of them focus on Fortune 500 Company CEOs, military and political leaders, athletic coaches, and religious leaders. However, the specific topic of construction leadership is rarely discussed. Even most of the academic articles with "construction leadership" in their titles do not really focus on the construction industry or construction personnel. Through direct interaction with over 300 design and construction professionals, the authors have developed the following definition of a construction leader as a person who:

- Guides others in the construction of a project through knowledge of work processes and thorough planning;
- Has influence over the construction process as a result of superior experience or authority and is responsible for its outcome;
- Creatively inspires and enables others to perform construction tasks; and
- Has the skills, abilities, authority, power, and capacity necessary to guide, motivate, and influence the craftsmen in a way that minimizes conflict and results in a successful construction project.

Our study of construction leadership resulted in the following conclusions:

- A lot of research has gone into which leadership style is the best or most appropriate. Many of the previous *leadership theories* have evolved into a situational model—basically, "it depends." There are many theories, concluding with situational and adaptive, as to the most appropriate leadership style in construction.
- Many of us have potential *leadership capabilities*, but it takes a combination of natural talent and acquired or learned skills to be an effective construction leader. Construction leaders need a variety of tools in their toolboxes, including the ability to be both a manager and a leader. A construction leader's potential requires some education and a lot of experience. A construction leader has learned their trade by "doing": they can walk-the-talk.
- Many *traits* are included in the profile of a construction leader, with trust, respect, experience, communication, and vision topping the list.
- Many of the *stereotypes* of construction "bosses" or leaders have (thankfully) evolved into project managers and superintendents who are sensitive about those they work for, including clients and architects, and those they supervise at the project level. Construction leaders today

defy many *myths*; they are not the largest white males on the job who can scream and curse better than others but comprise a diverse group of individuals who excel in computer skills and personal relationship development.

- No one can build a large building or infrastructure by themselves; it takes a cohesive group of qualified engineers, craftsmen, and subcontractors functioning as a *team*, and it is up to the PM and superintendent to focus all these individuals and companies toward one common goal.
- Respect and trust are thought of in construction with a *360-degree* view: you have to give it to receive it.
- Construction craftsmen are not *motivated* solely by a Friday paycheck; rather, they place high emphasis on participation and inclusion.
- Although many feel that contractors want to win at all costs (i.e., they are totally focused on profit), a satisfied client who receives their project on time, exceptional quality, and a safe experience for the craftsmen are important aspects of *construction winning* for true leaders.
- OICs develop project managers to become corporate officers, PMs develop PEs to become PMs, and superintendents develop foremen to become future superintendents. A *construction leader's legacy* is not only their completed construction projects, but the future leaders they have helped develop.
- The CEO of any company, in the construction industry or others, is by definition the company's leader. The project manager and superintendent of a construction company are the leaders on the jobsite.

Summary

The project manager's role in a construction project may slightly vary depending upon the firm's home-office organization and culture, the size and type of a particular construction project, and the experience of the project manager. Although, as we have shown, many construction management activities can be delegated to project engineers to provide them with growth opportunities, the ultimate success of the project falls on the shoulders of the project manager and the superintendent.

Review questions

1. Other than those listed above or in Chapter 19, list three subcontracts or purchase orders that you as the PM should author and three you would delegate to your PE.
2. List five steps the PM can perform to ensure prompt payment from the client.
3. Why should the PM take the lead and ensure prompt payment to his or her subcontractors and suppliers?
4. Why would you, as the PM, want to market to a client or design professional?
5. Who should draft the prime construction contract, and why him or her?
6. Who should execute or sign the prime construction contract, and why him or her?
7. Why, as the PM, do you want to take the lead on the development of the estimate and contract schedule?
8. As the PM, who should you request assistance from to develop (a) the estimate and (b) the schedule?
9. List at least five individuals you would include in your "inner-circle" if you were working on a project as (a) the GC's PM, (b) the superintendent, or (c) a PE.
10. What are the four aspects of project controls we discussed?

Exercises

1. Describe two distinct and different job functions for all these job titles:
   a. GC's CEO
   b. Project manager
   c. Superintendent
   d. Senior project engineer
   e. Project engineer
   f. Foreman
2. List five methods the project manager can use to successfully mentor his or her project engineers in order for them to one day become project managers as well.
3. List five methods you would use to market yourself and your firm to potential clients and designers.
4. Of the project controls we discussed, argue why one of them is the most critical and/or the least critical. Note: there is no "right" answer here.
5. We have all had good and bad bosses. Describe why one individual in your life was a particularly good or bad boss. If a bad boss, what could that boss have done to be better? What can you do to make yourself a good boss?
6. We listed only five leadership traits. Which one of these should be at the very top of the list? What are five additional leadership traits that should be added to the list?

Reference

Schaufelberger J. and Holm L. (2017). *Management of Construction Projects: A Constructor's Perspective* (2nd edition). Abingdon, Oxon, UK: Routledge.

APPENDIX A
Case studies

This appendix briefly describes two case studies, which are intended to provide examples of actual construction documentation to support the textbook narrative. Although these cases, including the construction companies, project owner, and designers are all fictional, they are representative of actual situations and professional documentation. Parts and pieces of several projects worked on by the authors have been assembled to form these case studies. Many of the figures used in the text connect to the cases and hopefully weave a continued thread for the reader, connecting all of the construction management concepts, particularly as they relate to the role of the project engineer. Some of the figures are abbreviated from complete examples that are included on the companion website. The cases we have included and will refer to throughout the text include:

Case study 1:   Highway overpass
Case study 2:   Mixed-use development

Case study 1: Highway overpass

This project involves the reconstruction of an existing highway overpass in Montana State on Interstate 90 (I-90) near Missoula, Montana. The current overpass had been there for 40 years, had served its useful purpose, and was in need of significant seismic repair and expansion to handle increased traffic flow. The state designed the project with in-house staff and opened it to a competitive public bid on a lump sum basis comprised of unit prices. The successful general contractor was Gateway Construction Company out of Spokane, Washington. The State of Montana's bid form had 200 separate standalone unit price bids. Gateway's total bid was $9,510,000 for a 14-month project. Their jobsite general conditions amounted to $912,060, which was approximately 9.6% of the total bid. Abbreviated examples of their bid form and general conditions estimate are included in Chapter 17. The contractor's detailed jobsite general conditions estimate is included on the companion website. Additional estimating backup documents included in Chapter 17 are a portion of Gateway's work breakdown structure (WBS), a detailed quantity take-off (QTO), and a detailed pricing recap sheet. Other examples from this case study are also included in Chapters 7, 8, 19, and 20 and discussed in exercises distributed throughout the chapters. The project's organization chart is attached as Figure A.1.

Case study 2: Mixed-use development

This project is a mixed-use development (MXD) that delivered a wood-framed apartment and retail facility for a private project owner. MXD projects are those that include more than one and often three or more different types of uses within the same facility. In this case, a developer built a four-story wood-framed 100-unit apartment building over one floor of cast-in-place concrete core and shell space, which would later be built out as a restaurant, coffee shop, grocery store, and

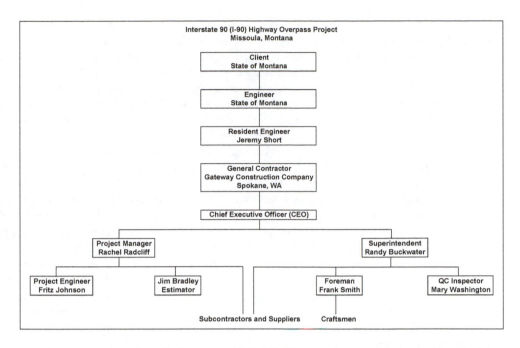

Figure A.1 Organization Chart for Interstate 90 (I-90) Highway Overpass Project, Missoula, Montana

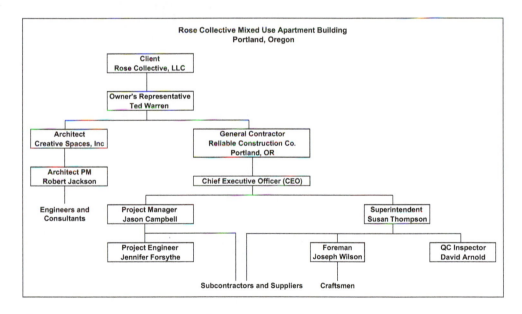

Figure A.2 Organization Chart for Rose Collective Mixed-Use Apartment Building Project, Portland, Oregon

some offices. Below this "podium" level were two stories of post-tension concrete parking garage, which would also serve as a separate profit center for the developer. This project was built in the Rose Quarter of Portland, Oregon, which is a thriving area for young educated professionals who are active in their academic, artistic, and athletic pursuits. The developer chose to name the project as the "Rose Collective" to represent a gathering place for the mix of different types of people in this new active generation, often labeled "Bohemian" Portland. Because of the dense neighborhood, parking was in short demand, so the developer decided to rent some of the new parking stalls to adjacent businesses. This developer had been active in the built environment for a long time and had many successful projects with repeat design and construction team members, so the project price was negotiated at 50% design completion. Three contractors were invited to prepare competitive proposals, and one was selected after interviews based upon a long list of criteria including fee, organization, budget, and past experience with mixed-use projects constructed in busy neighborhoods. The successful contractor was Reliable Construction Company (RCC), also out of Portland. The developer and Reliable executed an AIA A102 guaranteed maximum price (GMP) contract which included an 80–20 savings split. The total price was over $18 million. RCC's detailed construction estimate, their general conditions estimate, and their summary estimate are included on the companion website. An 18-month detailed construction schedule is also on the website. The GC's summary schedule and a typical three-week schedule are included in Chapter 18. Additional example documents for this project are included in Chapters 8, 9, 12, 14, 16, and 19 and discussed in exercises distributed throughout the chapters. This project's organization chart is included as Figure A.2.

APPENDIX B
Abbreviations

2D	two-dimensional
3D	three-dimensional
4D	four-dimensional
AB	anchor bolt
ACM	agency construction manager or management
ADA	Americans with Disabilities Act
ADR	alternative dispute resolution
AE	architectural, engineering, and related services
AEC	architectural, engineering, and construction
AHJ	authority having jurisdiction
AIA	American Institute of Architects
AMG	automated machine guidance
ASI	architect's supplemental instruction
ASPE	American Society of Professional Estimators
ATC	alternative technical concept
BDB	bridging design–build
BE	built environment
BEA	Bureau of Economic Analysis
BIM	building information modeling
BLS	Bureau of Labor Statistics
CAD	computer-aided design
CCA	construction change authorization
CCD	construction change directive
CD	change directive
CDs	construction documents
CDB	competitive design–build
CEO	chief executive officer
CF	cubic feet
CFL	compact fluorescent lamp (lighting)
CFO	chief financial officer
CII	Construction Industry Institute
CIM	civil information modeling
CIP	cast in place (concrete)
CM	construction manager or management
CMAA	Construction Management Association of America
CM/GC	construction manager as a general contractor
CMR	construction manager, or management, at risk
CO	change order

COBie	Construction Operations Building information exchange
C of O	certificate of occupancy
COO	chief operating officer
COP	change order proposal
CSI	Construction Specifications Institute
CY	cubic yard
DB	design–build (delivery method)
DBB	design–bid–build (delivery method)
DBFO	design, build, finance, and operate (delivery method)
DBIA	Design–Build Institute of America
DD	design development document
DFH	doors, frames, and door hardware
DRB	dispute resolution board
EA	each
EIA	U.S. Energy Information Administration
EJCDC	Engineers Joint Council Document Committee
EMR	experience modification rating (safety)
EPA	Environmental Protection Agency
EPC	engineering procurement and construction (delivery method)
ERP	enterprise resource planning
EV	earned value
EVM	earned value management
FICA	Federal Insurance Contributions Act (social security)
FO	field order
FWO	field work order
GA	Green Associate (LEED)
GBCI	Green Building Certification Institute
GBI	Green Building Institute
GC	general contractor
GCC	Gateway Construction Company (case study 1, fictitious contractor)
GC/CM	general contractor as a construction manager
GDP	gross domestic product
GHG	greenhouse gas (emissions)
GIS	geographic information systems
GLB	glue lam beam (engineered lumber)
GMP	guaranteed maximum price
GPS	global positioning system
GSA	General Service Administration
GWB	gypsum wall board
HCFC	hydrochlorofluorocarbon
HERS	Home Energy Rating System
HFC	hydrofluorocarbon
HM	hollow metal
HR	hour
HSS	hollow structural section
HUD	U.S. Department of Housing and Urban Development
HVAC	heating, ventilating, and air conditioning (subcontractor or mechanical system)
I-90	Interstate 90 (case study 2 project)
IFC	Industry Foundation Classes

InfraBIM	infrastructure BIM
IPD	integrated project delivery
IRS	Internal Revenue Service
ISI	Institute for Sustainable Infrastructure
ITB	instruction to bidders or invitation to bid
JHA	job hazard analysis
JV	joint venture
LBC	Living Building Challenge
LCI	Lean Construction Institute
LDs	liquidated damages
LED	light-emitting diode (lighting)
LEED	Leadership in Energy and Environmental Design
LF	linear feet
LLC	limited liability company
LSL	laminated strand lumber (engineered lumber)
LVL	laminated veneer lumber (engineered lumber)
MEP	mechanical, electrical, and plumbing (systems and contractors)
MH	man-hour
Mo or mos	month or months
MPC	multi-prime contracting
MSDS	material safety data sheet (now SDS)
MT	Montana State
MXD	mixed-use development
NA	not applicable
NAHB	National Association of Home Builders
NAICS	North American Industry Classification System
NASA	National Aeronautics and Space Administration
NCR	non-conformance report (similar to quality control)
NFPA	National Fire Protection Association
NGHBS	National Green Home Building Standard
NIBS	National Institute of Building Sciences
NIOSH	National Institute of Occupation Safety and Health
NTP	notice to proceed
O&M	operation and maintenance (manual)
OAC	owner–architect–contractor (coordination meeting)
OE	operating engineer
OEC	owner–engineer–contractor (coordination meeting)
OH&P	home-office overhead and profit (also known as fee)
OIC	officer-in-charge
OJT	on-the-job training
OR	Oregon State
OSB	oriented strand board (plywood)
OSHA	Occupational Safety and Health Administration
PDB	progressive design–build
PDF	portable document format
PDM	project delivery method
PE	project engineer
PEFC	Programme for the Endorsement of Forest Certification
PH	person-hour

PM	project manager or management
PMI	Project Management Institute
PO	purchase order
PPE	personal protection equipment (safety)
PR	proposal request
PSL	parallel strand lumber (engineered lumber)
PV	photovoltaic
QA	quality assurance
QC	quality control
QTO	quantity take-off
QTY	quantity
RCC	Reliable Construction Company (case study 2, fictitious contractor)
Rebar	concrete reinforcement steel
Recap	cost recapitulation sheet (estimating)
RESNET	residential energy services network
RFI	request for information or interpretation
RFP	request for proposal
RFQ	request for qualifications or request for quotation
R/I	rough-in
ROM	rough order of magnitude (budget estimate)
SDS	safety data sheet (formerly MSDS)
SF	square feet
SFCA	square feet of contact area
SFI	Sustainable Forestry Initiative
SK	sketch
SOG	slab on grade (concrete)
SOV	schedule of values
SWPPP	storm water pollution prevention plan
SY	square yard
T	ton or tonne or tonnage
TCO	temporary certificate of occupancy
TCY	truck cubic yards (earthwork)
TI	tenant improvement
TJI	Trus Joist International® (engineered lumber)
TQM	total quality management
UMH	unit man-hours (labor productivity estimating)
UN	United Nations
UP	unit price
U.S.	United States
USACE	United States Army Corp of Engineers
USGBC	United States Green Building Council
VDC	virtual design and construction
VE	value engineering
VECP	value engineering change proposal
WA	Washington State
WBS	work breakdown structure
Wks	weeks
WSDOT	Washington State Department of Transportation

APPENDIX C
Glossary

Active quality management program: process that anticipates and prevents quality control problems, rather than just responding to and correcting deficiencies

Activity duration: the estimated length of time required to complete an activity

Addenda: additions to or changes in bid documents issued prior to bid and contract award; also referred to as *amendments*

Agency construction manager: a firm or individual coordinating the work of prime contractors that contracts directly with the owner

Agreement: see *contractual agreement*

Allowance: an amount stated in the contract for inclusion in the contract sum to cover the cost of prescribed items, the full description of which is not known at the time of bidding. Once the architect or project owner provides a full description, the actual cost of such items is determined by the contractor and the total contract amount is adjusted accordingly

Alternatives: selected items of work for which bidders are required to provide prices

Alternative technical concepts: value engineering tool used predominantly in the public heavy-civil sector to promote innovation during procurement; deviations from the project concept, design, or specifications that are submitted by proposers to the project owner for their approval. Once approved, they can be incorporated into the final proposal, which provides a competitive advantage for the proposing contractor

Amendments: see *addenda*

American Institute of Architects: a national association that promotes the practice of architecture and publishes many standard contract forms used in the construction industry

Application for payment: see *payment request*

Apprentices: entry-level workers who are learning a trade through on-the-job training (OJT) under the supervision of other highly skilled workers, such as journey-level craftspeople

Arrow-diagramming method: scheduling technique that uses arrows to depict activities and nodes to depict events or dates

As-built drawings: contractor-corrected construction drawings depicting actual dimensions, elevations, and conditions of in-place constructed work

As-built estimate: assessment in which the actual costs incurred are applied to the quantities installed to develop actual unit prices and productivity rates

As-built schedule: marked-up, detailed schedule depicting actual start and completion dates, durations, deliveries, and restraint activities

Associated General Contractors of America: a national trade association primarily made up of construction firms and construction industry professionals

Authority having jurisdiction: "an organization, office, or individual responsible for enforcing the requirements of a code or standard, or for approving equipment, materials, an installation, or a procedure" (NFPA 2017)

Back charge: general contractor charge against a subcontractor for work that the general contractor performed on behalf of the subcontractor

Bar chart schedule: time-dependent schedule system without nodes that may or may not include restraint lines

Bid bond: a surety instrument that guarantees that the contractor, if awarded the contract, will enter into a binding contract for the price bid and provide all required bonds. A commonly used form is AIA document A310 that identifies "the maximum penal amount that may be due to the owner if the selected bidder fails to execute the contract or fails to provide any required performance and payment bonds" (AIA)

Bid peddling: unethical subcontractor's activity of offering to lower the bid provided to the general contractor on bid day; bid peddling "occurs when a subcontractor approaches a general contractor with the intent of voluntarily lowering the original price below the price level established on bid day. This action implies that the subcontractor's original price was either padded or incorrect" (ASPE 2011)

Bid shopping: unethical general contractor's activity of sharing subcontractor bid values with the subcontractor's competitors in order to drive down prices; bid shopping "occurs when a contractor contacts several subcontractors of the same discipline in an effort to reduce the previously quoted prices" (ASPE 2011)

Budget estimate: approximate cost estimate, usually developed early in the design process and also known as a *conceptual cost estimate* or *rough order of magnitude cost estimate*

Buildability: see *constructability*

Builder's risk insurance: protects the contractor in the event that the project is damaged or destroyed while under construction

Building information modeling: computer design software tools involving multi-dimensional overlays of building project information; a new approach toward the integration of design and construction processes for delivering facilities

Built environment: facilities and physical infrastructures that add or change functions in the underlying natural, economic, and social environments

Buyout: the process of awarding subcontracts and issuing purchase orders for materials and equipment

Buyout log: a project management document in spreadsheet format that is used for planning and tracking the buyout process

Cash flow curve: a plot of the estimated value of work to be completed each month during the construction of a project

Cash-loaded schedule: a schedule in which the value of each activity is distributed across the activity, and monthly costs are summed to produce a cash flow curve

Certificate of insurance: a document issued by an authorized representative of an insurance company stating the types, amounts, and effective dates of insurance for the designated insured

Certificate of occupancy: a certificate issued by the city or municipality indicating that the completed project has been inspected and meets all code requirements

Certificate of substantial completion: a certificate signed by the client, architect, and contractor indicating the date that substantial completion was achieved

Change order: modifications to contract documents made after contract award that incorporate changes in scope and adjustments in contract price and time. A commonly used form is AIA document G701

Change order proposal: a request for a change order submitted to the client by the contractor, or a proposed change sent to the contractor by the client requesting pricing data

Change order proposal log: a project management document in spreadsheet format listing all change order proposals and indicating dates of initiation, approval, and incorporation as final change orders

Civil information modeling: computer design software tools involving multi-dimensional overlays of infrastructure project information; a new approach toward the integration of design and construction processes for delivering infrastructures

Claim: an unresolved request for a change order

Close-out: the process of completing all construction and paperwork required to complete the project and close-out the contract

Close-out log: a project management document in spreadsheet format listing all close-out tasks that is used to manage project close-out

Commissioning: a process of testing and ensuring that all equipment is working properly

Conceptual cost estimate: cost estimates developed by using incomplete project documentation, also known as a *rough order of magnitude cost estimate* or *budget estimate*

Conceptual design: initial design developed during the planning phase as part of the conceptualization of the project idea and evolved during the initial stage of design activities; also known as *project concept*

ConsensusDocs®: family of contract documents which has taken the place of the AGC contract documents

Constructability: the ability of a designed project to be buildable as designed; also referred to as buildability

Constructability analysis: an evaluation of preferred and alternative materials and construction methods

Construction change directive: a directive issued by the client to the contractor to proceed with the described change order

Construction documents: construction documents take the design to a greater level of detail to include construction details and specifications for materials

Construction drawings: portion of the construction documents providing a graphical description of the project, including its geometrical information and often its materials

Construction manager: when referred to individuals, this term is used to identify all management positions on a project; when referred to firms, this term is used to identify firms that provide construction management services and serve a similar role to general contractors except they do not self-perform scope and, therefore, do not hire direct craft labor

Construction manager at risk delivery method: a delivery method in which the client has two contracts: one with the architect and one with the construction manager/general contractor. The general contractor is usually hired early in the design process to perform pre-construction services. Once the design is completed, the construction manager/general contractor constructs the project

Construction manager/general contractor delivery method: see construction manager at risk delivery method; also GC/CM

Construction specifications: portion of the construction documents that provide a textual description of the project, including any additional information on its materials, quality acceptance processes, or other performance expectations. Construction specifications are usually organized according to some standardized classification system. For instance, the Construction Specification Institute (CSI) MasterFormat and UniFormat are two commonly used classification systems for organizing data about construction requirements, products, and activities

Construction Specifications Institute: the professional organization that developed the original 16-division MasterFormat, which is the most used approach to organize technical specifications; the current version of the CSI MasterFormat includes 49 divisions

Construction surety: see *surety*

Contingencies: a portion of the estimating *markups* that takes into consideration potential inaccuracies due to uncertainties in the project definition and risks

Contract: a legally enforceable agreement between two parties; this is usually a written document, but some jurisdictions allow for oral contracts; see also *contractual agreement*

Contract schedule: also known as the formal schedule. These schedules will be provided to the client at the beginning and throughout the project delivery as required by the contract special conditions

Contract time: The period of time allotted in the contract documents for the contractor to achieve substantial completion; also known as *project time*

Contracting: see *project contracting*

Contractual agreement: a written document that sets forth the provisions, responsibilities, and the obligations of parties to a contract. Standard forms of agreement are available from professional organizations

Coordination drawings: multi-discipline design drawings that include overlays of mechanical, electrical, and plumbing systems with the goal of improving constructability and reducing change orders

Corrected estimate: estimate that is adjusted based on buyout costs

Cost codes: codes established in the firm's accounting system that are used for recording specific types of costs

Cost estimating: process of identifying the anticipated cost of a scope of work given the parameters available

Cost plus a fee contract: see *cost plus contract*

Cost plus contract: a cost-reimbursable contract in which the contractor is reimbursed for stipulated direct and indirect costs associated with the construction of a project and is paid a fee to cover profit and company overhead; also known as a *cost plus a fee contract*; see also *cost-reimbursable contract*

Cost plus contract with guaranteed maximum price: a cost plus contract in which the contractor agrees to bear any construction costs that exceed the guaranteed maximum price, unless the project scope of work is increased; see also *cost plus contract*

Cost plus with fixed-fee contract: a cost plus contract in which the contractor is guaranteed a fixed fee, irrespective of the actual construction costs; see also *cost plus contract*

Cost plus with incentive-fee contract: a cost plus contract in which the contractor's fee is based on measurable incentives, such as actual construction cost or construction time. Higher fees are paid for lower construction costs and shorter project durations; see also *cost plus contract*

Cost plus with percentage-fee contract: a cost plus contract in which the contractor's fee is a percentage of the actual construction costs; see also *cost plus contract*

Cost-reimbursable contract: a contract in which the contractor is reimbursed stipulated direct and indirect costs associated with the construction of a project. The contractor may or may not receive an additional fee to cover profit and company overhead

Craftspeople: non-managerial field labor force who construct the work, such as carpenters and electricians; see also *journey-level craftspeople*

Critical path: the sequence of activities on a network schedule that determines the overall project duration

Daily job diary: also known as the daily journal or daily report. A daily report prepared by the superintendent that documents important daily events including weather, visitors, work activities, deliveries, and any problems

Davis–Bacon wage rates: prevailing wage rates determined by the U.S. Department of Labor that must be met or exceeded by contractors and subcontractors on federally funded construction projects

Delivery: see *project delivery*

Design–bid–build delivery method: a delivery method in which the project owner hires a single contractor who constructs the project based on the design provided by the project owner; the design is substantially completed before a contractor is selected through a bid process

Design–build delivery method: a delivery method in which the client hires a single contractor who designs and constructs the project

Design–Build Institute of America: organization that defines, teaches, and promotes best practices in design–build

Design development: advanced design package outlining the specifications, including architectural information, such as floor plans, sections, and elevations, as well as layouts of the structural, mechanical, electrical, and plumbing systems

Detailed cost estimate: extensive estimate based on definitive design documents. Includes separate labor, material, equipment, and subcontractor quantities. Unit prices are applied to material quantity take-offs for every item of work. This type of estimate is most often associated with competitively lump sum bid projects and unit price bid projects

Differing site conditions: encountering site conditions different from those assumed in developing the design

Direct costs: costs directly relating to specific work items, for example, labor, materials, equipment, and subcontractor costs for the contractor, exclusive of any markups

Dispute: a contract claim between the owner and the general contractor that has not been resolved

Earned value: estimated value of the work completed to date

Earned value management: a technique for determining the estimated or budgeted value of the work completed to date and comparing it with the actual cost of the work completed. Used to determine the cost and schedule status of an activity or the entire project

Eighty–twenty rule: see *Pareto principle*

Enterprise resource planning (ERP) platform: company-wide software platform to integrate multiple business processes

Errors and omissions insurance: also known as professional liability insurance. Protects design professionals from financial loss resulting from claims for damages sustained by others as a result of negligence, errors, or omissions in the performance of professional services

Estimate schedule: management document used to plan and forecast the activities and durations associated with preparing the cost estimate. Not a construction schedule

Exhibits: important documents that are attached to a contract, such as a summary cost estimate, schedule, and document list

Expediting: process of monitoring and actively ensuring vendors' compliance with purchase order delivery requirements

Expediting log: a project management document in spreadsheet format used to track material delivery requirements and commitments

Experience modification rating: a factor that reflects the construction company's past claims history. This factor is used to increase or decrease the company's workers' compensation insurance premium rates

Fast-track construction: overlapping of design and construction activities, so that some are performed in parallel, rather than in series; allows construction to begin while the design is not yet complete

Feasibility analysis: analysis performed at the planning stage to evaluate the proposed project's ability to fulfill needs, as well as the technical and economic feasibility of the project concept

Fee: contractor's income after direct project and job site general conditions are subtracted. Generally includes home-office overhead costs and profit

Field engineer: project engineer responsible for field control and monitoring activities and usually less experienced than an office project engineer; may also be tasked with assisting the superintendent with technical office functions

Field question: see *request for information*

Field question log: see *request for information log*

Filing system: organized system for the storage and retrieval of project documents

Final completion: the stage of construction when all work required by the contract has been completed

Final inspection: final review of project by client and architect to determine whether final completion has been achieved

Final lien release: a lien release issued by the contractor to the client or by a subcontractor to the general contractor at the completion of a project to indicate that all payments have been made and that no liens will be placed on the completed project

Fixed price contract: see *lump-sum contract*

Float: the flexibility available to schedule activities that are not on the critical path without delaying the overall completion of the project

Foreman: direct supervisor of craft labor on a project; on larger projects, more than one individual may have the title of foreman and serve under a leading foreman who is titled *general foreman*

Free on board: a material item whose quoted price includes delivery at the point specified. Any additional shipping costs are to be paid by the purchaser of the item; also known as freight on board

Front loading: a tactic used by a contractor to place an artificially high value on early activities in the schedule of values to improve cash flow

General conditions (as a document): one of the contract documents, which describes operating procedures that the owner usually uses on all projects of a similar type. The conditions describe the relationship between the owner and the contractor, the authority of the client's representatives or agents, and the terms of the contract. The general conditions contained in AIA document A201 are used by many clients

General conditions (as costs): indirect costs, whether in the home office or at the jobsite, that cannot be attributed solely to any direct work activities

General contractor: construction firms with knowledge of the whole process to provide overall supervision and coordination through employment of direct craft labor and subcontractors. They usually serve as prime contractor and agree to construct the project in accordance with the contract documents while being responsible for the overall planning and coordination of construction operations

General foreman: see *foreman*

General liability insurance: protects the contractor against claims from a third party for bodily injury or property damage

General superintendent: see *superintendent*

Geotechnical report: also known as a *soils report*; a report prepared by a geotechnical engineering firm that includes the results of soil borings or test pits and recommends systems and procedures for foundations, roads, and excavation work

Great Recession: a period of significant slowdown in economic activity with reduction in the amount of goods and services produced and sold that occurred in the late 2000s and strongly affected the economic viability of built environment firms and employment in related industries

Guaranteed maximum price contract: a type of *open-book* or *cost plus contract* in which the contractor agrees to construct the project at or below a specified cost and potentially share in any cost savings; a guaranteed maximum price can be added to a cost plus contract later on

Hazardous materials: construction materials, either part of the final project or used in means and methods, that may cause damage to people or the environment if not handled properly; part of the active safety plan

Heavy-civil contractor: term sometimes used to distinguish general contractors that only build infrastructures. They usually self-perform more tasks and often own a fleet of construction equipment

Home-office indirect costs: portion of indirect costs that cannot specifically apply to the contract work but represents the cost of doing business; also referred to as *home-office overhead*

Home-office overhead: see *home-office indirect costs*

Indirect construction costs: expenses, whether in the home office or at the jobsite, indirectly incurred and not directly related to a specific project or construction activity, such as *home-office overhead* or *jobsite overhead*

Initial inspection: a quality-control inspection to ensure that workmanship and dimensional requirements are satisfactory

Insurance company: a company using analyses to evaluate the chances an event may occur and its financial impact on a project and to establish a premium price for the insurance company taking on that risk

Integrated project delivery: "a project delivery approach that integrates people, systems, business structures and practices into a process that collaboratively harnesses the talents and insights of all participants to optimize project results, increase value to the owner, reduce waste and maximize efficiency through all phases of design, fabrication and construction" (AIA)

Interoperability: ability to communicate externally with other computer tools; a crucial aspect to consider when selecting a BIM platform or evaluating the project owner's requirements on BIM

Invitation to bid: a portion of the bidding documents soliciting bids for a project; also known as instructions to bidders

Job hazard analysis: the process of identifying all hazards associated with a construction operation and selecting measures for eliminating, reducing, or responding to the hazards

Jobsite overhead: field indirect costs that cannot be tied to an item of work but are project-specific and, in the case of cost-reimbursable contracts, are considered part of the cost of the work; also referred to as *general conditions* (as costs)

Joint venture: a contractual collaboration of two or more parties to undertake a project

Journey-level craftspeople: see *craftspeople*

Just-in-time delivery of materials: a material management philosophy in which supplies are delivered to the jobsite just in time to support construction activities; this minimizes the amount of space needed for on-site storage of materials

Labor and material payment bond: see *payment bond*

Land surveyors: licensed professionals performing construction surveys to identify the position of the proposed facility or infrastructure and other surveys, such as site utilities and topography mapping

Lean Construction: "Lean Construction extends from the objectives of a Lean production system—maximize value and minimize waste—to specific techniques, and applies them in a new project delivery process" (Seed 2010); a process to improve costs by incorporating efficient methods during both design and construction; includes value engineering and pull planning

LEED: Leadership in Energy and Environmental Design, a measure of sustainability administered by the United States Green Building Council; usually associated with receipt of a LEED certificate

Letter of intent: a letter, in lieu of a contract, notifying the contractor that the client intends to enter into a contract pending resolution of some restraining factors, such as permits, financing, or design completion; sometimes allows limited construction or procurement activities to occur

Lien: a legal encumbrance against real or financial property for work, material, or services rendered to add value to that property

Lien release: a document signed by a subcontractor or the general contractor releasing its rights to place a lien on the project

Life-cycle cost: the sum of all acquisition, operation, maintenance, use, and disposal costs for a product over its useful life cycle

Liquidated damages: an amount specified in the contract that is owed by the contractor to the client as compensation for damages; the value incurred as a result of the contractor's failure to complete the project by the date specified in the contract

Litigation: a court process for resolving disputes

Long-form purchase order: a contract for the acquisition of materials that is used by the project manager or the construction firm's purchasing department to procure major materials for a project; see also *purchase order*

Look-ahead schedule: see *short-interval schedule*

Lump-sum contract: A contract that provides a specific price for a defined scope of work; also lump sum bid or estimate; also known as a *fixed-price contract* or *stipulated-sum contract*

Markup: price items that are not actual costs, but refer to the reward of doing business in the form of profit and the risk of adverse circumstances in the form of insurance and contingencies; often a percentage added to the cost of the work to cover such items as home-office overhead, taxes, and insurance and to provide for a profit and contingency funds (if not specifically forbidden by contract terms)

MasterFormat: a 16-division numerical system of organization developed by the Construction Specifications Institute that is used to organize contract specifications and cost estimates

Materialman's notice: a notice sent to the client as notice that the supplier will be delivering materials to the project

Material safety data sheet: see *safety data sheet*

Material supplier: vendor who provides materials but no on-site craft labor

Meeting agenda: a sequential listing of topics to be addressed in a meeting

Meeting notes: a written record of meeting attendees, topics addressed, decisions made, open issues, and responsibilities for open issues

Meeting notice: a written announcement of a meeting. It generally contains the date, time, and location of the meeting, as well as the topics to be addressed

Mock-ups: standalone samples of completed work, such as a 10-foot-by-10-foot sample of an architecturally finished concrete wall

Network diagrams: schedule that shows the relationships among the project activities with a series of nodes and connecting lines

Notice to proceed: written communication issued by the owner to the contractor that authorizes the contractor to proceed with the project and establishes the date for project commencement

Occupational Safety and Health Administration: federal agency responsible for establishing jobsite safety standards and enforcing them through inspection of construction work sites

Officer-in-charge: general contractor's principal individual who supervises the project manager and is responsible for overall contract compliance; also *project executive*

Off-site construction: off-site pre-fabrication of building modules or systems

Open-book accounting: payment method requiring the construction firm to show its costs to the other party; it provides transparent access to cost information to the purchasing party but adds more bureaucracy to the contract administration

Operation and maintenance manuals: a collection of descriptive data needed by the client to operate and maintain equipment installed on a project

Overbilling: requesting payment for work that has not been completed

Overhead: expenses incurred that do not directly relate to a specific project, for example, rent on the contractor's home office; see also *general conditions* (as costs) and *home-office indirect costs*

Overhead burden: a percentage markup that is applied to the total estimated direct cost of a project to cover overhead or indirect costs

Pareto principle: on most projects, 20% of the invested input is responsible for 80% of the results obtained; about 80% of the costs or schedule durations are included in 20% of the work items; also referred to as the 80–20 rule or *eighty–twenty rule*

Payment bond: a surety instrument that guarantees that the contractor (or subcontractor) will make payments to their craftspeople, subcontractors, and suppliers. A commonly used form is AIA document A312; also known as *labor and material payment bond*

Payment request: document or package of documents requesting progress payments for work performed during the period covered by the request, usually monthly; also referred to as *application for payment*

Performance bond: a surety instrument that guarantees that the contractor will complete the project in accordance with the contract. It protects the client from the general contractor's default and the general contractor from the subcontractor's default. A commonly used form is AIA document A312

Plugs: general contractor's early cost estimates for subcontracted scopes of work

Post-project analysis: reviewing all aspects of the completed project to determine lessons that can be applied to future projects

Pre-bid conference: meeting of bidding contractors with the project client and architect. The purpose of the meeting is to explain the project and bid process and to solicit questions regarding the design or contract requirements

Precedence-diagramming method: scheduling technique that uses nodes to depict activities and arrows to depict relationships among the activities; used by most scheduling software

Pre-construction agreement: a short contract that describes the contractor's responsibilities and compensation for pre-construction services

Pre-construction conference: meeting conducted by client or designer to introduce project participants and to discuss project issues and management procedures

Pre-construction services: services that a construction contractor performs for a project client during design development and before construction starts

Pre-final inspection: an inspection conducted when the project is near completion to identify all work that needs to be completed or corrected before the project can be considered completed; a form of active quality control

Preparatory inspection: a quality-control inspection to ensure that all preliminary work has been completed on a project site before starting the next phase of work

Pre-proposal conference: meeting of potential contractors with the project client and architect. The purpose of the meeting is to explain the project, negotiating process, and selection criteria and to solicit questions regarding the design or contract requirements

Pre-qualification of contractors: investigating and evaluating prospective contractors based on selected criteria prior to inviting them to submit bids or proposals

Prime contract: agreements between the project owner and another party

Prime contractors: construction firms that contract directly with the project owner, in opposition to *subcontractors* that do not have a direct contractual relationship with the project owner

Procurement: see *project procurement*

Product data sheet: also known as a material data sheet or cut sheet. Information furnished by a manufacturer to illustrate a material, product, or system for some portion of the project; it can include illustrations, standard schedules, performance data, instructions, and warranty

Profit: the contractor's net income after all expenses have been subtracted

Progress payments: periodic (usually monthly) payments made during the course of a construction project to cover the value of work satisfactorily completed during the previous period

Project close-out: see *close-out*

Project concept: see *conceptual design*

Project contracting: the process of establishing a contractual relationship for services and/ or materials through the development of a written agreement expressing the expectations, responsibilities, and protections of each party. Contracting for construction services is substantially different from contracting for design services

Project control: methods the project team utilizes to anticipate, monitor, and adjust to risks and trends in controlling costs, schedules, quality, and safety

Project delivery: the act of achieving project objectives

Project delivery method: "defines the relationships, roles, and responsibilities of project team members and the sequence of activities required to complete a project" (Gibson and Walewski 2001)

Project finance method: a system for acquiring or providing funds from different sources and combining them for financing a project during its delivery

Project engineer: project management team member responsible for management of technical issues on the jobsite who assists the project manager and superintendent on larger projects; project engineers may have different levels of experience and responsibilities, from field engineering all the way to assisting the project manager

Project executive: general contractor's principal individual who supervises the project manager and is responsible for overall contract compliance; also *officer-in-charge*

Project labor curve: a plot of estimated labor hours required per month for the duration of the project

Project management: application of knowledge, skills, tools, and techniques to the many activities necessary to complete a project successfully

Project manager: the leader of the contractor's project team who is responsible for ensuring that all contract requirements are achieved safely and within the desired budget and time frame; usually supervises field office staff, including project engineers

Project manual: a specification volume that may also contain contract documents, such as the instructions to bidders, bid form, general conditions, special conditions, and/or the geotechnical report

Project planning: the process of selecting the construction methods and the sequence of work to be used on a project

Project procurement: the act of purchasing external services and materials necessary to deliver a project

Project-specific safety plan: a detailed accident prevention plan that is focused directly on the hazards that will exist on a specific project and on measures that can be taken to reduce the likelihood of accidents

Project superintendent: see *superintendent*

Project team: individuals from one or several organizations who work together as a cohesive team to construct a project

Project time: see *contract time*

Public private partnership: a public agency partners with a contractor or developer, in the case of construction, to reduce costs and lawsuits and, ultimately, to save the taxpayer money; also known as P3

Pull Planning: scheduling method, often utilizing sticky notes, where the milestones of each design or construction discipline are established and the project is scheduled backward with the aid of short-term detailed schedules; a tool of *lean construction*

Punch list: a list of items that need to be corrected or completed before the project can be considered completed

Purchase orders: written contracts for the purchase of materials and equipment from suppliers; see also *long-form purchase order* and *short-form purchase order*

Quality control: process to ensure materials and installations meet or exceed the requirements of the contract documents

Quantity take-off: one of the first steps in the estimating process to measure and count items of work to which unit prices will later be applied to determine a project cost estimate

Reimbursable costs: costs incurred on a project that are reimbursed by the client. The categories of costs that are reimbursable are specifically stated in the contract agreement

Renewable materials: those materials that can be reproduced at a rate that meets or exceeds the rate of human consumption

Request for information: document used to clarify discrepancies between differing contract documents and between assumed and actual field conditions; also known as a *field question*

Request for information log: a project management document in spreadsheet format for tracking requests for information from initiation through designer response

Request for proposals: document containing instructions to prospective contractors regarding documentation required and the process to be used in selecting the contractor for a project

Request for qualifications: request for prospective contractors or subcontractors to submit a specific set of documents to demonstrate the firm's qualifications for a specific project

Request for quotation: a request for prospective contractors to submit a quotation for a defined scope of work

Retention: a portion withheld from progress payments for contractors and subcontractors to create an account for finishing the work of any parties not able to or unwilling to do so; also known as retainage

Risk management: method used to understand project risks and either accept them, mitigate them, transfer them to other parties, or insure against them

Rough order of magnitude cost estimate: a *conceptual cost estimate*, usually based on the size of the project. It is prepared early in the estimating process to establish a preliminary budget and decide whether or not to pursue the project

Safety data sheet: short technical report that identifies all known hazards associated with a particular material and provides procedures for using, handling, and storing the material safely; formerly a *material safety data sheet*

Schedule of submittals: a project management document in spreadsheet format that lists all submittals required by the contract specifications

Schedule of values: an allocation of the entire project cost to each of the various work packages required to complete the project. Used to develop a cash flow curve for an owner and to support requests for progress payments; serves as the basis for AIA document G703, which is used to justify pay requests

Schedule update: schedule revision to reflect the actual time spent on each activity to date

Schematic design: design expanding the project scope, which can be used as the basis for later design development

Self-performed work: project work performed by the general contractor's workforce, rather than by a subcontractor

Shell and core project: building project with scope limited to building structure, envelope, site work, and building systems

Shop drawing: drawing prepared by a contractor, subcontractor, vendor, or manufacturer to illustrate construction materials, dimensions, installation, or other information relating to the incorporation of the items into a construction project

Short-form purchase order: purchase orders used on project sites by superintendents to order materials from local suppliers; see also *purchase order*

Short-interval schedule: schedule that lists the activities to be completed during a short interval (2–4 weeks). Used by the superintendent and foremen to manage the work; also known as a *look-ahead schedule*

Site logistics plan: pre-project planning tool, often created by the general contractor's superintendent, which incorporates elements including temporary storm water control, hoisting locations, parking, trailer locations, fences, traffic plans, etc.

Soils report: see *geotechnical report*

Special conditions: a contract document that describes operating procedures that are unique requirements for the project. It is used to supplement and/or modify, add to, or delete portions of the *general conditions* (as document); also known as *supplementary conditions*

Specialty contractors: construction firms that specialize in specific areas of construction work, such as painting, roofing, or mechanical systems. Specialty contractors usually, but not necessarily, participate in a project as subcontractors. However, they may also serve as *prime contractors* under multi-prime contracting, renovation, and tenant improvement projects

Stipulated sum contract: see *lump-sum contract*

Subcontract agreements: see *subcontracts*

Subcontracting plan: document prepared by the general contractor once a *work-breakdown structure* is ready. It identifies which scopes to include in each subcontract, essentially dividing up the pie among each intended subcontract

Subcontractor: construction firms that do not have a direct contractual relationship with the project owner, in opposition to *prime contractors* that contract directly with the project owner. Subcontractors are usually, but not necessarily, specialty contractors that contract with and are under the supervision of a prime contractor that is usually, but not necessarily, a general contractor

Subcontractor call sheet: a form used to list all of the bidding firms from which the general contractor is soliciting subcontractor and vendor quotations

Subcontractor plugs: see *plugs*

Subcontractor pre-construction meeting: a meeting the project manager and/or superintendent conduct with each subcontractor before allowing them to start work on a project; the client and architect may also be invited

Subcontracts: all contracts following the hierarchical structure of the project delivery method that do not involve the project owner as one of the parties; written contracts between the general contractor and specialty contractors who provide jobsite craft labor and, usually, material for specialized areas of work; also known as *subcontract agreements*

Submittal: document or product generated by the construction team to verify that what they plan to purchase, fabricate, deliver, and ultimately install is in fact what the design team intended by their drawings and specifications; examples of submittals include shop drawings, product data sheets, and samples submitted by contractors and subcontractors for verification by the design team that the materials intended to be purchased for installation comply with the design intent

Submittal log: see *schedule of submittals*

Substance abuse: use of illegal drugs or alcohol, either prior to coming to work on a construction site or during work, which could impair the performance of an individual and place him or her and others around them in jeopardy of being hurt; dealt with as part of an active safety plan

Substantial completion: state of a project when it is sufficiently completed that the owner can use it for its intended purpose

Summary schedule: abbreviated version of a detailed construction schedule that may include 20–30 major activities

Superintendent: individual from the contractor's project team who is the leader on the jobsite and who is responsible for supervision of daily field operations on the project; on larger projects, more than one individual may have the title of superintendent and serve under a leading superintendent who is titled *general superintendent* or *project superintendent*

Supplementary conditions: see *special conditions*

Suppliers: firms providing only material and equipment for a project and supporting the installation by others of the material or equipment they provide

Surety: company able to provide various guarantees on behalf of another company in the form of bonds. A surety investigates the project and its participants to evaluate if the chances of a risky event occurring are low enough to issue a bond. Three types of bonds are prevalent in the industry: *bid bond*, *performance bond*, and *payment bond*

Sustainability: broad term incorporating many green building design and construction goals and processes, including LEED

Sustainable project: built environment project incorporating some sustainable concepts among its project objectives and aiming to produce a facility or infrastructure that is able to strike a balanced fit with its underlying social, economic, and natural environments

Technical specifications: a part of the construction contract that provides the qualitative requirements for a project in terms of materials, equipment, and workmanship

Tenant improvement project: building project with scope limited to finishes, partitions, and trimming of mechanical, electrical, and plumbing systems

Third-tier subcontractor: a subcontractor who is hired by a firm that has a subcontract with the general contractor

Time-and-materials contract: a *cost plus contract* in which the client and the contractor agree to a labor rate that includes the contractor's profit and overhead. Reimbursement to the contractor is made based on the actual costs for materials and the agreed labor rate times the number of hours worked

Total quality management: a management philosophy that focuses on continual process improvement and customer satisfaction

Trade damage: damage to a finished work item by individuals other than those employed by the contractor that has completed the damaged work item

Traditional project delivery method: a delivery method in which the client has a contract with an architect to prepare a design for a project. When the design is completed, the client hires a contractor to construct the project

Transmittal: a form used as a cover sheet for formally transmitting documents between parties

Unit price contract: a contract that contains an estimated quantity for each element of work and a unit price; the actual cost is determined once the work is completed and the total quantity of work is measured; often associated with heavy-civil public works bid projects, such as a highway

Value engineering: a methodical process to evaluate a project's design in order to obtain the best value for the cost of construction; at the detail level, it results in a series of studies of the relative value of various materials and construction techniques to identify the least costly alternative without sacrificing quality or performance

Value engineering change proposal: a request for a change order submitted to the client by the contractor that would allow implementation of value engineering through cost saving or value adding

Warranty: guarantee that all materials furnished are new and able to perform as specified and that all work is free from defects in material or workmanship

Work breakdown structure: "a hierarchical decomposition of the total scope of work to be carried out by the project team to accomplish the project objectives" (PMI); a list of significant work items that will have associated cost or schedule implications

Workers' compensation insurance: insurance that protects the contractor from a claim due to injury or death of an employee on the project site

Work package: a defined segment of the work required to complete a project

Index

WITHDRAWN

CPSIA information can be obtained
at www.ICGtesting.com
Printed in the USA
LVHW061148200119
604573LV00010B/178/P

9 781138 736580